Copper for America

COPPER FOR AMERICA

The United States Copper Industry from Colonial Times to the 1990s

Charles K. Hyde

The University of Arizona Press
Tucson

The University of Arizona Press
www.uapress.arizona.edu

 Published 1998
First paperback edition 2016

Printed in the United States of America
21 20 19 18 17 16 7 6 5 4 3 2

ISBN-13: 978-0-8165-1817-3 (cloth)
ISBN-13: 978-0-8165-3279-7 (paper)

Cover design by Miriam Warren
Cover photo: Plant and smokestack of a large copper smelter of the Phelps-Dodge Mining Company at Morenci, Arizona, 1942. Photo by Fritz Henle, Office of War Information, courtesy of the Library of Congress, LC-USE6- D-010043 [P&P].

Publication of this book is made possible in part by a grant from Wayne State University.

Library of Congress Cataloging-in-Publication Data
Hyde, Charles K., 1945–
Copper for America : the United States copper industry from colonial times to the 1990s / Charles K. Hyde.
p. cm.
Includes bibliographical references and index.
ISBN 0-8165-1817-3 (acid-free paper)
1. Copper mines and mining—United States—History. 2. Copper industry and trade—United States—History. I. Title.
TN443.A5H93 1998
338.2'743'0973—dc21
97-45308

British Library Cataloguing-in-Publication Data
A catalogue record for this book is available from the British Library.

♾ This paper meets the requirements of ANSI/NISO Z39.48-1992 (Permanence of Paper).

To Henry C. Hyde Sr.,

scrap metal dealer,

for teaching me the value of the red metal

Contents

Maps

Tables

Preface

The intellectual roots of this book extend back to 1977, when I conducted an inventory of historic engineering and industrial sites in the Upper Peninsula of Michigan for the Historic American Engineering Record (HAER), an agency of the National Park Service. The impressive physical remains of the copper mines of the Keweenaw Peninsula first ignited my interest in the copper industry. I returned to the Lake Superior mining district in the summer of 1978 with a HAER recording team headed by Larry Lankton to document the physical remains of the Quincy mine. Larry Lankton and I subsequently produced *Old Reliable: An Illustrated History of the Quincy Mining Company* (1982), though we both continued to study the copper industry, we followed separate trails. Larry recently published *Cradle to Grave: Life, Work, and Death at the Lake Superior Copper Mines* (1991), a well-received "social history of technology" of the Michigan copper mining industry. This book, in contrast, examines the history of the American copper industry from its colonial roots to the recent past. It is an attempt to integrate the economic and business history of the copper industry with its technological, social, and labor histories. Throughout the book, both major and minor copper-producing districts are examined in detail, with appropriate comparisons and contrasts. For the period from about 1840 to 1920, the histories of the principal producing districts serve as the organizational framework for the book.

In the course of more than a decade of research, I have used the resources of dozens of archives and have tried the patience of dozens of archivists. I have received enormous assistance from Theresa Sanderson Spence of Michigan Technological University's Copper Country Historical Collections. Leroy Barnett of the State Archives of Michigan and Peter L. Steere of the Special Collections at the University of

Arizona Library were especially generous with their time as well. The following archival institutions also gave me courteous and prompt service over the course of my research: the Bentley Historical Collections, University of Michigan; the Maryland Historical Society (Baltimore); the Historical Society of Pennsylvania (Philadelphia); the American Philosophical Society (Philadelphia); the Montana Historical Society (Helena); the Anaconda Historical Society (Anaconda, Mont.); the Arizona Historical Society (Tucson); the Manuscript Department of the Arizona State University Library (Tempe); the Arizona Historical Foundation (Tempe); Cline Library, Northern Arizona University (Flagstaff); the Arizona Division, Arizona Department of Library, Archives, and Public Records (Phoenix); and the Huntington Library (San Marino, Calif.). In addition, Collamer M. Abbott of White River Junction, Vt., provided me with a copy of his unpublished book manuscript on the Vermont copper industry.

I am particularly grateful to the generous friends and colleagues who read parts of the book manuscript: Alan R. Raucher of Wayne State University; Michael P. Malone of Montana State University; Robert L. Spude of the Rocky Mountain (Denver) Office of the National Park Service; and Fredric L. Quivik. Wayne State University has supported my work with two sabbatical leaves of absence. Of course, none of these individuals or institutions are culpable for any errors or omissions found in this book.

Introduction

Historians have by no means ignored the history of the American copper industry, particularly the colorful history of its prospectors, miners, and capitalists. While copper cannot claim the glamour of the precious metals or the significance of coal or iron ore to the national economy, it has remained a key metal in industrial economies everywhere. Mining engineer Thomas Rickard devoted more than one-third of his classic *History of American Mining* (1932) to the red metal. Historians of western mining, however, have often treated copper simply as one of a half-dozen metals exploited in the course of the opening of the region, and a poor relation of gold and silver. More recent work on western mining has focused on mine workers in all the branches of the industry, with copper miners given scant attention.[1]

Scholars have produced valuable studies of the copper districts of Michigan and Montana, which include their economic and business histories but also have serious limitations. William B. Gates's *Michigan Copper and Boston Dollars* (1951) is more than forty years old, while Michael P. Malone's *The Battle for Butte* (1981) considers Montana's copper industry only through 1906. No similar surveys exist for Arizona or the rest of the West. Recent books on labor relations in the Michigan, Montana, and Arizona mining districts provide useful information on the copper industry and its workers. In addition, professional and amateur historians have published more than fifty substantial biographies and company histories over the past half-century, but these are of uneven quality and value.[2]

Many of the general histories of the copper industry published in recent decades have been less than comprehensive in scope. Several have treated the operation of the American and international copper markets and the changing global copper industry broadly but give little attention to the business history of the major American copper companies. There are, however, two major exceptions. Thomas R.

Navin's *Copper Mining and Management* (1978) examines the internal business histories of the principal American copper companies and discusses international developments, including the nationalization of American-owned mines overseas. Second, George H. Hildebrand and Garth L. Mangum's *Capital and Labor in American Copper, 1845–1990* (1992) is an ambitious effort to integrate developments in the American copper industry's product and labor markets over the industry's entire history.[3]

Is an additional book on the copper industry of the United States needed? All of the general histories discussed above begin with the development of the Michigan copper district in the mid-1840s, as if the United States had no copper industry before then. The first chapter of this book treats the copper industry during the colonial period and in the early Republic within the context of domestic economic policy and the global market. Significant mines in Vermont, Massachusetts, Connecticut, New Jersey, Maryland, Pennsylvania, and Tennessee yielded a substantial quantity of the red metal before the Michigan mines opened in the 1840s. In addition, copper rolling and smelting industries developed at various times between 1801 and the early 1850s in Boston and Taunton, Massachusetts; Baltimore, Maryland; Belleville, New Jersey; and Ducktown, Tennessee.

This book focuses on the development of the three principal American copper mining districts: Michigan, Montana, and Arizona. Chapters 2 and 3 consider the long history of the Michigan district, the first major copper mining region in the United States. The Montana copper industry is considered in Chapter 4, while Chapters 5 and 6 examine the Arizona mining district and the copper industry in the rest of the West. A regional approach makes less sense after 1920, when a small number of national and multinational copper companies have dominated the industry, and the regional copper industries have become less distinct. The final two chapters trace the history of the industry from 1920 to the late 1980s.

In analyzing the development of the Michigan, Montana, and Arizona mining districts, I examine the economic and business histories of the producing companies, including the careers of their leaders; the evolving technologies of mining, concentrating, smelting, and refining; and the development of global copper production and consumption. For each of the major mining districts, I analyze the impact of the varying forces that influenced the development of the copper industry. None of the published regional studies have achieved this degree of integration and synthesis.

More important, I compare and contrast the character and development of the three principal copper mining districts. They each have distinctive histories, in terms of the initial discoveries of copper, the sources of investment capital used to bring the mines into production, and the character of the mine owners and managers. These and other contrasts between the three principal copper districts are especially interesting and instructive.

While using a regional framework to understand much of the copper industry's history, this book also strives to integrate developments at the regional level with changes in the national and global copper industries. The copper producers in the individual districts produced for both the national and global markets. American copper companies cooperated to gain tariff protection after the Civil War and to establish international cartels beginning in the late 1880s. After several large American copper companies—including Anaconda, ASARCO, and Kennecott—invested heavily in overseas mines starting in the 1910s, the U.S. copper industry became genuinely multinational in character.

The last two chapters of this book attempt to explain the long-term decline of the American copper industry since the 1910s, when the United States produced more than half of the world's copper. Much of this decline seems inevitable, given the existence of vast undeveloped copper deposits in other parts of the globe and the predictable exhaustion of proven deposits within the United States. In some respects, however, the decline was the result of conscious decisions about investment priorities. Starting in the 1910s, the largest American producers, particularly Anaconda and Kennecott, began to invest heavily in overseas mines while ignoring their domestic properties. They failed to modernize their facilities and to develop improved technologies that might have allowed them to profitably exploit lower grades of ore. They also failed to spend much of their capital in exploration efforts. Instead, they invested in other industries or simply distributed their profits as dividends. In short, there is strong evidence that managerial shortsightedness and incompetence contributed mightily to the industry's decline in the twentieth century.

Copper for America

CHAPTER 1

Foundations

Copper is an article so necessary to us at present for sheathing ships, for making distilling vessels, . . . for coin, etc., etc., that I hardly know of any manufacture of such importance, after iron: and yet we have no smelting work for copper, or any copper mine worked in the United States.—Thomas Cooper, in *The Emporium of Arts and Sciences*, June 1814

The Europeans who explored and settled the North American continent hoped to repeat the success of their Spanish counterparts to the south in gaining instant wealth by expropriating or mining precious metals and gems. The Jamestown settlers, for example, came equipped with mining tools, and their numbers included goldsmiths and refiners. After about four years of "gold fever," they finally turned to agriculture and discovered the value of tobacco as a cash crop. Ironically, the Virginia colony survived in part by mining the deposits of iron ore found in the area. To be sure, both the French and the English turned to the fur trade as an alternative source of treasure, especially as the prospects for finding precious metals faded. While searching for gold and silver explorers also looked for evidence of base metals, including copper.[1]

No sustained copper mining occurred in the colonies, however, until the first decade of the eighteenth century. One of the earliest successful colonial copper mines was developed at Simsbury (now East Granby), Connecticut, midway between Hartford and Springfield, Massachusetts. The Simsbury mine produced small amounts of copper spasmodically between roughly 1720 and 1788. The Connecticut colony purchased the property in 1773 and converted it into Newgate Prison, incarcerating prisoners underground. Some convicts mined and smelted ore at Newgate until 1788, when the colony permanently suspended mining. This facility housed Tory sympathizers during the American Revolution, then served as Connecticut's state prison until 1827, when a new prison opened.[2]

The Middle Atlantic colonies, particularly New Jersey, produced the largest volumes of copper ore throughout the colonial period. The Schuyler mine, located about four miles northeast of downtown Newark, was easily the most famous and most productive of the New Jersey copper mines. It derived its name from Arent Schuyler (1662–1730), the landowner who first developed the deposits. He was shipping copper ore to England by at least 1715, but following his death in 1730, productivity and profits declined.[3]

Josiah Hornblower and John Stearndall leased the property for fourteen years starting in 1761. During Hornblower's tenure the Schuyler mine experienced a series of disasters that ultimately produced its permanent closing. The engine house burned in March 1762, in July 1768, and finally in 1773. Hornblower rebuilt the engine and reopened the mine following the 1762 and 1768 disasters, but it was never again profitable. His account books suggest production of ore ranging between a low of about ten tons in 1765 and 1772 to a peak of about forty tons in 1771. Following the last engine-house fire, Hornblower abandoned the property.[4]

The success of the Schuyler mine spawned many other copper mining ventures in New Jersey and elsewhere in the Middle Atlantic colonies. Substantial failed copper mines included the Rockey Hill mine in Franklin, thirty miles southwest of Newark; the French mine near New Brunswick; the Bridgewater mine near Somerville, New Jersey; the Mineral Hill mine in Carroll County, Maryland; and the Liberty mine in Frederick County, Maryland.[5] Virtually every colonial copper mine had to rely on foreign workers: at Simsbury, Germans developed the mine and built a stamp mill and smelter; the Schuyler mine used Cornish miners starting in 1754 and hired the Cornish engineer Josiah Hornblower to erect a steam engine and to manage the mine; Frind Lucas, an Englishman, managed the Dod

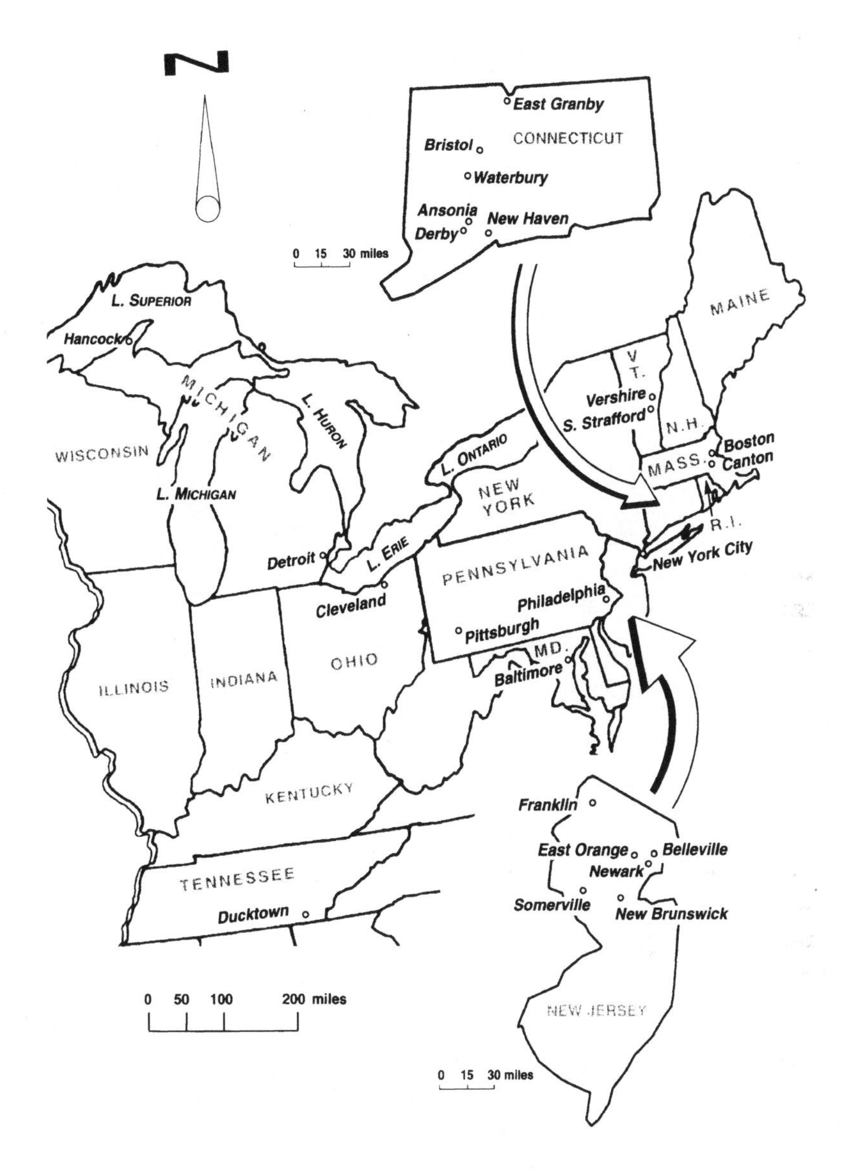

Mining, smelting, and fabricating sites before the Civil War.
(Map by Mike Brooks)

mine in East Orange, New Jersey, from 1739 until around 1760; the Rockey Hill mine in New Jersey employed German miners, hired a German metallurgist, J. F. Reynier, in 1755, and when Rockey Hill reopened in 1765, it employed 160 Welsh miners; finally, English miners developed the Mineral Hill mine and the Fountain Copper Works in Maryland, with stamp mills and a smelter.[6]

The first serious effort to mine Lake Superior copper occurred in the early 1770s. Alexander Baxter, Henry Bostwick, and Jean Baptiste Cadotte, based in Sault Sainte Marie, established a partnership to explore for copper and received a royal charter in 1770 for their corporation, the Proprietors of Mines on Lake Superior. They built a forty-ton sloop at the Sault, hired Alexander Henry as a guide, and in the spring of 1771, sent a party of Cornish miners to the Ontonagon River to open a mine. They worked there for a year, managed to drive a forty-foot tunnel into the side of a mountain, but they found no copper and abandoned the venture in the spring of 1772. The onset of the American Revolution, disputes between the British and American governments over control of the Great Lakes, and the War of 1812 discouraged further exploration for half a century.[7]

Assessing the colonial copper industry in quantitative terms is difficult, given the lack of reliable statistics on output. For most mines, evidence about production is fragmentary or nonexistent. The American colonies may have produced their largest output (perhaps between a hundred and two hundred tons of metallic copper annually) in the 1720s and 1730s, when the Schuyler and Simsbury mines were at their peak production. Copper from new mines developed in New Jersey and Maryland in the 1750s probably did not make up for the decline of the Schuyler and Simsbury mines. Colonial copper production was probably well under a hundred tons annually by the 1750s, and on the eve of the American Revolution it was no more than a trickle. Exports of copper ore in 1770, for instance, amounted to only forty-one tons.[8]

Ironically, American demand for copper and brass grew rapidly beginning around 1750, just as the domestic copper industry began a long period of decline. The colonial population—still under a half-million in 1720—had more than doubled by 1750 and reached 2.8 million in 1780. British foreign-trade statistics show the substantial growth of American imports of copper and brass. The copper equivalent of imports from Britain fluctuated between 10 and 20 tons per annum in the first quarter of the eighteenth century, but then averaged between 40 and 50 tons over the two decades ending in 1745. At mid-century, imports stood at 151 tons and rose to a peak of 350 tons in 1760. Over the period 1750–1770, the American colonies took between one-tenth and one-fifth of British copper and brass exports.[9]

Next to iron, copper and brass were the most useful metals in the colonial economy. Colonial coppersmiths crafted a bewildering variety of practical and decorative products, with domestic and industrial kettles, pots and pans, and stills the most important. The number of coppersmiths working in colonial America, and their concentration in a handful of New England and Middle Atlantic cities, is evident from Henry Kauffman's documented listing of 81 coppersmiths who worked before 1790. Boston and Philadelphia each had 19, followed by Baltimore (11), eastern Pennsylvania (11), New York City (9), and southern New England (7).[10]

Brass founders also produced a wide variety of products, including andirons, bells, cannon, small arms, and molds for pewter castings. Kauffman identified 46 brass founders who worked before 1790, also concentrated in a handful of cities. Philadelphia led with 19, followed by New York City (12), Boston (4), Baltimore (3), and Annapolis (3). In describing Philadelphia's industries, Leander Bishop noted that "Brass-founding and copper, brass, and tin work of all kinds for distilleries, brewing, sugar-mills in the West Indies, and refineries of sugar, and for household use, employed many tradesmen in Philadelphia from an early period." The demand for copper and brass continued to grow through the rest of the eighteenth century and contributed to a revival of mining as the century ended.[11]

Copper in the Early Republic

American consumption of copper and brass grew considerably between about 1790 and 1840, during a time when the United States had no significant domestic source of copper. American customers remained a small but growing part of a rapidly expanding world demand for copper, primarily supplied by Cornish mines and by the newly emerging mining region of Chile. During the first two decades of the nineteenth century, world copper production averaged between 9,000 and 9,500 tons per annum. Following the Napoleonic Wars and the brief but severe economic depression following the panic of 1817, global output boomed in the 1820s, averaging 13,500 tons yearly, and in the 1830s reached 21,800 tons per annum. Britain typically produced between two-thirds and four-fifths of world production in 1790–1840 and routinely exported between half and three-quarters of its domestic output.[12]

Cornwall's mines and Swansea's smelters dominated the world copper industry through the 1830s. The volume of ore mined in Cornwall grew impressively from about 50,000 tons in 1798 to over 100,000 tons by 1822 and 150,000 tons by the early 1840s. Ore production grew more rapidly than metal output, however, because the copper content of Cornish ores declined over time. Cornish ores mined in 1799 yielded 10 percent copper but averaged about 9 percent in 1800–1819 and only 8 percent in the 1820s and 1830s. Merely maintaining a constant level of metallic copper production required more ore. Between 1799 and 1840, for example, production of metallic copper from Cornish ores rose a healthy 124 percent, but ore tonnage jumped by 187 percent. Had the ores of 1840 had the metallic content of 1799, Cornish mines could have produced the 1840 volume of metal with only 110,000 tons of ore, rather than the actual output of 147,000 tons.[13]

The Swansea smelters used a technology originally developed in Germany in the sixteenth century. The smelters would roast the ore to drive out sulfur and then melt the resulting product to generate a slag, which they then drew off. Initially, the process involved dozens of repetitions and could take more than a year to complete. The Swansea smelters developed a simpler technique around 1750, the so-called Welsh Process, which by the 1840s consisted of only six stages and could be completed in about four days. At each stage the sulfur content of the material fell and the copper content increased. By the end of the fourth stage, the metal was already 70 percent copper, and by the end of the sixth stage, close to 100 percent. Final production of the refined copper took place in reverberatory furnaces, which allowed the flames from the fire to melt and refine the metallic copper but kept the fuel separated from the metal.[14]

The Welsh Process fit the conditions in South Wales well. It required enormous quantities of coal (about twenty tons per ton of copper) but less labor than methods used on the Continent. Coal was plentiful and cheap in the Swansea area, while wages were high. The smelter operators, particularly the Vivian family at its Hafod smelter, made substantial technological improvements. Between 1809 and 1842 the Vivians reduced the time needed for smelting by two-thirds and labor costs by half.[15]

Smelting technology encouraged the growth of large-scale integrated firms. The Welsh Process required a mixture of ores of various composition to ensure the most efficient smelting and the highest-quality copper. Impurities in the ores acted as reducing agents, or fluxes, for other ores, allowing the production of slag without the use of additional materials. Smelters achieved optimum efficiency when they could draw upon large stockpiles of ores. Additionally, the use of specialized furnaces for the various phases of smelting and refining made economic sense only if production was high.[16]

The beginning of the end of British hegemony in world copper markets came in the mid-1820s. English and Cornish capitalists invested heavily in Chilean copper mines during the speculative boom of 1823–25 and sent Cornish miners and engineers to develop the mines there. Initially, Chile's mines sent most of their ore to Swansea, but a domestic smelting industry emerged in Chile in the late 1820s. Chileans used smelting furnaces to produce "matte copper," a partially refined material containing between 40 and 60 percent metal, and crude reverberatory furnaces to convert the matte to metallic copper. By 1832, Chile had a total of forty-three smelting furnaces to produce matte copper and eleven reverberatory furnaces. By the late 1830s, Chile was exporting the equivalent of 13,200 tons of metallic copper annually, half in the form of bars, while the rest went as ore, primarily to Swansea.[17]

No American copper mines of any significance were at work until the late 1830s, except for the Schuyler mine, which operated between 1793 and 1806. The revival of the American copper industry began in the manufacturing stage of production, using foreign metal and ore. Demand for copper sheets rose sharply in the last quarter of the eighteenth century, with much of the metal used to make "flat and raised bottoms" for stills and vats, and for architectural applications, such as roofing, flashing, and gutters.[18] The earliest buildings roofed with copper included New York's city hall (1763), the Maryland statehouse (1773), Mount Vernon (1784), lighthouse towers in the 1790s, the building for the First Bank of the United States (1797) in Philadelphia, and the Massachusetts statehouse (1802). Copper flashing, gutters, and downspouts also became common by the end of the century.[19]

The use of copper to sheath ship bottoms starting in the late eighteenth century created an enormous new demand for copper. The hulls of wooden ships sailing in tropical waters faced three hazards: the teredo shipworm, which bored into the timbers; barnacles, which grew on ship bottoms and sides, reducing speed and maneuverability; and rotting, which the other problems accelerated. The British Navy introduced copper sheathing in 1761 and equipped most of the fleet with copper bottoms over a three-year period starting in 1779. The British merchant fleet also adopted copper sheathing in the 1780s, with the East India Company and slave traders leading the way.

American shipbuilders and owners did not lag far behind. One of the first American ships fitted with a copper bottom was the *Empress of China,* built in 1784 at Baltimore.[20] The U.S. Naval Act of 1794 authorized the construction of six American frigates, which the Navy Department agreed to outfit with copper bottoms. For a while, American metal merchants imported enough copper sheets to satisfy the requirements of the new navy. However, the undeclared naval war with France (1798–1800) and the wars with the Barbary states (1801–1805) led to a naval construction program that required vast amounts of copper. In 1798, Secretary of the Navy Benjamin Stoddert proposed the building of twelve seventy-four-gun ships, each requiring about 1,620 tons of copper bolts, spikes, and sheathing, which cost about $28,400 per ship for copper alone. In 1798, however, the British government prohibited the export of copper sheets suitable for sheathing, which produced a serious shortage of copper sheathing in the United States.[21]

As a direct result of this crisis, Paul Revere of Boston, better known as a silversmith and patriot, built the first rolling mill for copper in the United States. In 1797 he supplied the fledgling U.S. Navy with copper bolts and a bell for the *Constitution,* which was built a short distance from his foundry. The next year, Revere offered to supply the navy with brass cannon, bells, and copper bolts, but he also suggested that he could produce copper sheathing if he could find an adequate supply of good copper. Early in 1800, Revere won a contract for 20,000 pounds of sheathing. He risked about $25,000 of his own capital, and the government loaned him $10,000 and 19,000 pounds of copper. Revere bought an abandoned gunpowder mill on the Neponset River in Canton, Massachusetts, about sixteen miles from Boston, purchased rolls for his mill in England, and on 24 October 1801 rolled his first copper sheets.[22]

Revere produced sheathing that the navy deemed acceptable, but the federal government was so slow to pay him that he nearly went bankrupt. He survived by receiving prompt payment in 1802 for 8,000 pounds of sheathing and nails he provided for the dome of the new Massachusetts statehouse. When the *Constitution* returned to Boston in 1803 for refitting, Revere provided the sheathing, which the workers installed in only two weeks' time. He constantly faced shortages of copper and eventually persuaded the navy to have its ships carry copper as ballast on return voyages to the United States. Revere's rolling mill struggled in peacetime, when navy contracts dried up, but the Canton mill was hard at work again during the War of 1812, when Revere produced three tons of manufactured copper per week, primarily in the form of sheathing for ships.[23]

The extraordinary economic conditions created by the Napoleonic Wars brought two additional rolling mills into operation. Levi Hollingsworth, a prominent Baltimore merchant, started a small rolling mill in that city in 1804 or 1805, mainly to provide Baltimore's shipbuilding industry with sheathing, spikes, and rods. In 1814, using machinery imported from England, Hollingsworth opened a large copper manufacturing complex on the Gunpowder River some eleven miles north of the Baltimore city limits. The Gunpowder Copper Works included two reverberatory furnaces to refine bar and pig copper, a cupola furnace to process slag, and two sets of sheet rolls in a water-powered mill. It supplied the copper sheathing for the rebuilding of the Capitol dome in Washington in 1815. Hollingsworth ran this venture successfully until his death in 1822, when Isaac McKim managed it for the Hollingsworth family. The Gunpowder Works operated from 1814 to 1861 and again from 1866 to 1883 before closing.[24]

A third major American rolling mill went into operation in 1814. Harmon Hendricks (1771–1838), who had supplied sheathing to commercial vessels since 1801, bought the Soho smelter and foundry at Belleville, New Jersey, in May 1814. The Soho works, originally developed by Nicholas Roosevelt to smelt the Schuyler mine ore and to manufacture steam engines, had been idle since 1806. Hendricks formed a partnership with his brother-in-law, Solomon I. Isaacs, to build a rolling mill. The firm used the awkward name of S. I. Isaacs & Soho Copper Company until its dissolution in 1827, when it became the Soho Copper Company. The partnership built a new refinery and a well-equipped rolling mill, powered by the reliable Second River, and in October, 1814, the Soho works, also known as the Belleville or the Bergen Point works, rolled its first copper sheets. The Bergen Point works ran steadily into the 1840s, refining copper and rolling copper sheets.[25]

Federal tariff policies after 1790 encouraged copper and brass manufacturing but did not aid mining or smelting. The tariff enacted in June 1794 raised duties on manufactured copper and brass wares from 5 percent ad valorem to 15 percent, among the highest rates levied on any product. The standard duties on manufactured copper and brass products increased to 20 percent in 1816 and two years later jumped again to 25 percent, where they remained into the 1830s. Ingot and scrap copper remained duty-free until 1846, and copper sheathing paid no duties until the Civil War.[26]

American copper manufacturers benefited from a boom in American shipbuilding in the 1820s but suffered from cyclical downturns as well. In the *Digest of Manufactures* prepared for Congress in 1822, for example, the Soho works reported that it was operating at only half its capacity of 175 tons of copper and earning no profits. Revere's Canton mill, with a capacity of 200 tons, did not submit any information about production, suggesting that it was not operating. Despite volatile demand for copper sheets, Hendricks added a second rolling mill at Soho in 1824, increasing annual capacity to 350 tons.[27]

The growth in demand for copper continued through the 1820s, when Hendricks was the premier supplier of copper sheathing in the United States. The New York shipbuilding industry alone was recoppering at least one ship per week using his sheathing. In 1825–27, the Soho Works rolled over 1,000 tons of copper. Although copper mining languished through the 1820s, copper and brass manufacturing thrived despite the lack of significant tariff protection for rolling mill operators such as Revere, Hendricks, and Hollingsworth.[28]

During the 1830s the demand for copper began to shift from maritime uses into new applications. The Hendricks firm sold increasing amounts of the red metal to railroad locomotive manufacturers, who needed copper for boiler plates, tubes, flues, and boxes. During the peak sales period 1834–36, Hendricks sold $6,358 worth of copper to the U.S. Navy but more than $18,000 worth to Mathias W. Baldwin and to the West Point Foundry Association, both railroad locomotive manufacturers. In 1836 the Hendricks brothers invested $17,000 to improve the machinery at their Soho Copper Works, not anticipating the prolonged business crisis that would begin the following year.[29]

The three major copper rolling mill operators, the Reveres, Hollingsworth, and Hendricks, faced increased competition in the late 1820s and 1830s from new producers. Isaac McKim, who managed Levi Hollingsworth's Gunpowder Copper Works following Hollingsworth's death in 1822, built a rolling mill in 1827 on Smith's Wharf in Baltimore. The new mill, known as the Vallona Copper Works, made ship sheathing and other naval copper. McKim's mill became a tourist attraction because of its giant steam engine, which drove all the rolling equipment. The Baltimore & Cuba Smelting and Mining Company, founded by Haslett McKim (son of Isaac McKim) and Dr. David Keener in 1845, subsequently bought the Vallona mill.[30]

Another Baltimore manufacturer, E. T. Ellicott, periodically rolled copper through the 1820s at his iron works near Ellicott City, west of Baltimore. As late as 1826, Ellicott was supplying copper sheathing to the U. S. Navy. He apparently gave up the copper business and reverted to rolling iron in 1828, failing to find sufficient copper or profits to continue. By the late 1820s, Baltimore had become an important center for copper imports, in part because of the area firms rolling copper. The construction of copper smelters in Baltimore starting in 1845 further enhanced the city's strategic place in the American copper industry.[31]

With the virtual disappearance of the domestic copper mining industry between 1790 and 1830, copper and brass consumed in the United States came from foreign sources. Tracing the growth of domestic consumption is difficult because no reliable foreign-trade statistics exist. The United States government did not record the amount of goods imported duty-free until 1821, and importers systematically undervalued goods that were assessed ad valorem duties. The weaknesses in the foreign-trade figures are especially troublesome for scholars studying the copper industry because the overwhelming majority of imports were duty-free, including pigs, bars, and sheets of copper and brass; metal destined for the U.S. Mint; and copper used to make stills. However, the remaining manufactures of copper and brass were subject to the highest ad valorem rates on imports. Before 1821, estimates of consumption and trade must come from other sources.[32]

Circumstantial evidence suggests that copper and brass manufacturing enjoyed substantial growth between 1790 and 1820. Kauffman documented 81 coppersmiths working before 1790 and an additional 133 who began work between 1790 and 1820. Similarly, an additional 81 brass founders began working between 1790 and 1820, compared to the 46 who began work before 1790. More than half of the new brass founders were in Philadelphia.[33]

In the early 1790s, most copper went into ship sheathing, kettles, stills, vats, and brass wares. Lathrop estimated that American consumption of copper and brass combined was only about 150 to 200 tons in 1820, on the eve of the development of the Connecticut brass industry. His estimates are far too low. For 1821, the first year for which we have comprehensive trade figures, the total value of all copper and brass imports was about $650,000. Sheathing copper and old copper sold for about $600 and $300 per ton, respectively, in the same year. An average price for copper and brass imports of $450 a ton suggests imports of 1,400 tons, seven times Lathrop's figures. Exports and re-exports were nil.[34]

TABLE 1.1

Copper and Brass Imports, By Value, 1825–1840

Year Ending 30 Sept.	Copper Imports	Brass Imports	Total
1825	$734,615	$526,093	$1,260,708
1827	864,363	489,024	1,353,387
1830	807,238	373,273	1,180,551
1835	1,302,640	423,796	1,726,436
1840	1,653,315	249,354	1,902,669

Source: New American State Papers: Commerce and Navigation (Wilmington, Del.: Scholarly Resources, 1973), 10:26–27, 12:16–19, 15:12–13, 18:12–13, 22:12–13.

The volume of American foreign trade in copper cannot be precisely determined through 1824 because tin, copper, and brass are lumped into a single category. Imports of "copper, brass, and tin, in pigs or bars," however, grew dramatically from slightly less than $250,000 in 1821 to between $500,000 and $600,000 per annum in 1822–24, when this single category accounted for half of the combined imports of copper and brass. The import figures, summarized in table 1.1, show the growth of copper and brass consumption between 1825 and 1840, and the declining share of brass imports in the total. As late as 1840, the total value of all exports of American copper and brass was only $87,000, while re-exports remained negligible.[35]

Lathrop estimated total American consumption of copper and brass in the early 1830s at about 500 tons annually, but the trade statistics again suggest a much higher figure. Sheathing sold for $440 a ton in 1830, and old copper for $320 the same year. If the average price of all copper and brass imports, including manufactured products, was $400 a ton in 1830, imports were about 3,000 tons, six times Lathrop's estimate. By 1840, imports probably stood at about 4,000 tons.[36]

The detailed import figures reflect the changing patterns of copper consumption, as delineated in table 1.2. The most striking shift in the mix of imported copper was the rapid growth in imports of pig and bar copper, while imports of manufactured copper fluctuated erratically. Much of the larger volume of raw copper went to the

TABLE 1.2

Copper Imports, By Major Categories, By Value, 1825–1840

Year Ending 30 Sept.	Pigs & Bars	Bolts and Sheathing	Copper Bottoms & Vessels	Old Copper	Misc. Mfres.
1825	$199,354	$455,138	$53,207	—	$26,916
1827	183,080	441,767	34,513	$66,985	138,018
1830	417,635	286,188	4,847	83,413	15,198
1835	683,812	514,067	3,613	96,911	86,712
1840	1,100,664	412,999	8,809	70,405	60,438

Source: New American State Papers: Commerce and Navigation (Wilmington, Del.: Scholarly Resources, 1973), 10:12–13, 36–37, 80–81; 12:30–31, 40–41, 84–85; 15:25–33, 72–73; 18:36–37, 44–45, 84–85; 22:34–35, 42–43, 82–83.

expanding copper rolling industry and, starting in the late 1830s, to the Connecticut brass industry.

The U.S. Treasury Department's annual reports on American foreign trade show how the sources of copper imports into the United States changed over time. In 1825, England supplied virtually all the copper sheathing and other manufactured copper, and accounted for 77 percent of all copper imports, by value, or a total of $567,378. As early as 1825, Chile supplied nearly $110,000 of bars and pigs, or about 15 percent of the value of American copper imports. England was still providing the bulk of manufactured copper imports in 1840, but English imports comprised only a third of the total of about $1,650,000. In contrast, Chile and Peru provided more than $1 million in copper in 1840, 60 percent of the total, virtually all in the form of pigs and bars.[37]

The Revival of the Industry

In the two decades before the Civil War, copper mining enjoyed a rebirth, copper and brass manufacturing grew rapidly, and a substantial smelting industry emerged starting in the mid-1840s. This revival occurred while the global copper industry underwent a dramatic transformation. World metallic copper output, which stood at

40,000 tons in 1840, rose to 60,000 tons in 1850, and then reached 95,000 tons on the eve of the Civil War. The locus of the world copper industry, however, shifted dramatically from Great Britain to the New World. Metallic copper production from British ores increased from 13,000 tons in 1840 to 14,700 tons ten years later and in 1860 reached 16,000 tons. Britain still accounted for one-third of world output in 1840, but only one-quarter of the total in 1850, and one-sixth by 1860.[38]

There were few efforts to mine copper in the United States between about 1780 and the early 1830s, but speculators and serious investors then opened mines in Vermont, Connecticut, Maryland, New Jersey, and Tennessee. A substantial copper smelting industry emerged in the mid-1840s, based in part on ores from the new mining districts but primarily dependent on South American ores. Historians of the early Republic seldom acknowledge the emergence of an American copper industry in the eastern states starting in the 1830s. Instead, the enormous success of the Michigan copper district starting in the late 1840s has overshadowed the important advances made in previous decades.[39]

The east central region of Vermont seems like an unlikely site for a revival of American copper mining. A group of Boston capitalists who joined forces in 1809 to develop the iron sulphide deposits in South Strafford (Orange County), Vermont, to manufacture copperas, an iron sulphate, discovered that the deposits also contained chalcopyrite (copper pyrite). Isaac Tyson Jr., who held an 1827 patent for making copperas, and Dr. Amis Binney speculated in mining properties in Strafford and nearby Vershire in the early 1830s. Tyson managed a copperas works and copper smelter at Strafford in the 1830s. He experimented with anthracite coal and a hot blast for the smelting furnaces. In March 1832, *Niles' Register* reported that the copper smelter had a capacity of a half ton of metallic copper per day, but actual production is unknown. Copper mining and smelting continued at Strafford until 1839, when the company closed the operation.[40]

Mining languished in this district until the Vermont Copper Mining Company, chartered by the state legislature in November 1853, reopened the mine at Vershire. Known as the Copperfield mine but renamed the Ely in the 1870s, it ran continuously until 1883. The Copperfield mine produced a total of 3,270 tons of ore between 1854 and 1860, with output reaching 1,452 tons in 1860. During the Civil War, it produced between 1,500 and 1,800 tons of ore annually.[41]

Connecticut was also the scene of serious efforts to mine copper, with a mine at Bristol southeast of Hartford as the most important venture. In 1836, George W. Bartholomew removed large samples of ore, which he sent to England for smelting. Much of the ore was reportedly rich enough (up to 80 percent copper) to go into the smelting furnace without further concentration. In 1837 he launched the Bristol Mining Company. However, in 1845 control of the company passed into the hands of two New York investors, Richard H. Blydenburgh and Hezekiah Bradford, who also managed real estate ventures for Union College and its president, Dr. Eliphalet Nott. When Robert W. Lowber replaced Blydenburgh and Bradford as the financial agent for Union College in 1850, Nott received the Bristol mine as part of the financial settlement. Nott bought out Blydenburgh's interest in the mine for $31,000, loaned the mining company $212,052 in return for a mortgage on the property, and by January 1851 was the principal owner.[42]

Nott later hired two well-known geologists—Benjamin Silliman Jr. of Yale College and Josiah D. Whitney, the State Mineralogist of Iowa and an expert on copper—to examine his property. They produced a report offering a glowing view of the mine's prospects. Trusting Silliman and Whitney's judgment, John M. Woolsey acquired a majority interest in the Bristol Mining Company, which he reorganized in 1855, but he could not avoid declaring bankruptcy in 1858. Because of prolonged legal disputes over Woolsey's estate, the mine remained closed for thirty years. The losses at the Bristol mine were staggering even by mid-nineteenth-century standards. According to one estimate, investors had spent $870,000 on the venture before the company's final collapse.[43]

Maryland was another center of the revival of American copper mining in the 1830s, and Isaac Tyson Jr. (1792–1861) was the driving force behind most of the copper mines. Tyson, a Baltimore native, discovered chromite deposits near his hometown in 1827 and proceeded to buy all the significant chromite deposits of Pennsylvania and Maryland, enabling him to corner the world market for two decades. He then manufactured copperas and, as a result of his work in Strafford, Vermont, became interested in copper mines.[44]

Tyson eventually became involved in virtually all the Maryland copper mines developed before the Civil War. Among others, he operated the New London mine (1836–53), the Dolly Hide mine (1846–53), the Mineral Hill mine (1849–60), and the Springfield mine (1849–60). Tyson had a knack for making sick mines well again, at least for a time. With Tyson's successful mines operating in the 1840s and early 1850s, Maryland was easily the leading copper mining area in the United States before Michigan yielded vast amounts of copper in the 1850s. Tyson's impact as a geologist, chemist, engineer, and promoter on the development of mining and metallurgy in the eastern United States was substantial.[45]

New Jersey had more copper mining ventures than Maryland in the antebellum period, but virtually all of them were dismal failures. There were several unsuccessful attempts in the 1830s and 1840s to operate the famous Schuyler mine, closed since about 1806. The Bridgewater mine, also known as the American and as Van Horne's mine, located about ten miles northwest of New Brunswick, had produced copper until around 1775. The mine operated spasmodically between 1825 and the Panic of 1837, which served as its death knell. Other copper mining ventures included the Washington mine (1837–46) and the Flemington mine (1840–60).[46]

In describing copper mines throughout the United States in 1854, J. D. Whitney singled out the New Jersey mines for derision. Despite warnings from the New Jersey state geologist that most copper deposits in the state were not rich enough to merit investment, capitalists and speculators continued to pour money into them. Whitney noted that New Jersey had its own little copper investment "fever" in 1846–47, much like the one that struck the Lake Superior copper district, with six new mining companies launched in 1847 alone. He listed the major mine failures that resulted from foolish speculation, noting that "all these mines were abandoned, after heavy expenditures, with almost total loss of the whole amount invested; and it is to be hoped that no more money will be sunk in them."[47]

The last mining region to emerge before the Civil War, outside of

Michigan, was the Ducktown district of southeast Tennessee. Located immediately north of the Georgia border, the area derived its name from a Cherokee village, which had a resident chief named "Duck." No significant development occurred until 1847, when a German miner, A. J. Weaver (also known as Webber), shipped ninety casks of ore, more than 15 tons, to the Point Shirley smelter in Boston, which assayed two lots at 14.5 percent and 32.5 percent copper. Four mines were at work in the Ducktown district by the end of 1852, and a total of fourteen by August 1854. During three years of work ending in September 1853, the Ducktown mines shipped some 14,291 tons of ore to Swansea or east coast smelters, yielding about 3,500 tons of metallic copper. During the year ending July 1854, the district shipped about 4,200 tons, with an average copper content of 32 percent. S. W. McCallie claimed that in 1855 the district produced over 800 tons of ore *per month*.[48]

The first ore shipments from the Ducktown district caused unexpected problems. Damp chalcocite in casks was subject to spontaneous combustion because the ore contained sulfur. When an American vessel that sailed from Savannah with a mixed cargo of Ducktown ore and cotton arrived in Liverpool in June 1855, dock workers found charred wooden casks and partially burned cotton. Roasting the ore to drive out sulfur became an absolute necessity for safety reasons alone. Given the poor transportation system linking Ducktown to the outside and a tendency for the copper content of the ore to decline at greater depths, economics dictated that the mines ship their ore as regulus or matte copper, roughly 30 percent to 50 percent metal, or in the form of bars or ingots.[49]

The Panic of 1857 touched off a major consolidation of the Ducktown mines. In January 1858 a group of eastern and southern investors took over the Union Consolidated Mining Company of Tennessee and bought eleven mines, with a total of 2,575 acres of mineral rights. Julius E. Raht, a German mining captain who had managed several Ducktown mines, became general superintendent of the new company. The defunct Polk County Copper Company also reorganized in 1859 and named Raht as its mine superintendent. Finally, in April 1860 the newly organized Burra Copper Company of Tennessee purchased the Hiwassee, Cocheco, and Toccoee mines. The Burra

Copper Company, with Raht as superintendent, quickly built a major new smelter to treat its ores.[50]

By the onset of the Civil War, Ducktown production had already begun to decline as the mining companies exhausted the richest deposits. The Tennessee Rolling Works Company opened a rolling mill in Cleveland, Tennessee, thirty miles from Ducktown, just as the war began, so the disruption of trade with eastern (Yankee) markets had little impact on the Ducktown district. The smelting works concentrated on producing ingot copper, and the Confederacy purchased the district's entire output. The Ducktown mines remained at work and provided the South with much-needed copper until the fall of 1863, when Union forces destroyed the Cleveland rolling mill and guerilla warfare in east Tennessee made mining impossible.[51]

During the 1840s and 1850s, the American economy started down the road to self-sufficiency in copper. A large and vigorous domestic brass industry emerged in the Naugatuck River Valley in Connecticut, in part because of tariff protection against foreign (British) brass. A substantial smelting industry concentrated in a few east coast cities also emerged in the mid-1840s, using cheap South American ores, Pennsylvania anthracite coal, and increasingly, the output from an expanding domestic mining industry. These developments, along with the emergence of the Michigan mines in the 1850s, drastically reduced America's dependence on British copper.

The Connecticut brass industry was insignificant until the late 1820s. Aaron Benedict, who had manufactured buttons in Waterbury since 1812, and Israel Coe established the firm of Benedict & Coe in 1829 and built a larger brass rolling mill in Waterbury. In 1830, Holmes & Hotchkiss, another new firm producing brass sheets and wire, began operations in Waterbury. Israel Coe established the Wolcottville Brass Company in 1834 in partnership with John Hungerford and Anson G. Phelps, a New York City metal merchant. Anson Greene Phelps (1781–1853) grew up in Simsbury, Connecticut, and briefly worked as a saddler's apprentice in Hartford, but in 1812 he established himself as a merchant in New York City . In 1821, Phelps and Elisha Peck formed the partnership of Phelps and Peck, which specialized in imports of metals and metal manufactures, including copper, brass, tin, lead, and iron. Phelps dissolved his partnership with Peck in 1834, and with his son-in-law, William E. Dodge, founded Phelps, Dodge & Company. The new firm prospered in the 1830s largely by selling English tin plate.[52]

The success of the Wolcottville Brass Company encouraged Anson G. Phelps to increase his investments in the Connecticut brass industry. He built a brass rolling mill at Derby, seventeen miles below Waterbury, in 1836. In 1844, Phelps built a larger mill two miles upstream from Derby at a settlement he named Ansonia. A decade later, Phelps enlarged his mill at Ansonia and abandoned the factory at Derby. Another significant newcomer to the brass industry, the Waterbury Brass Company, incorporated in 1845 and built its first factory the following year. The emergence of the Connecticut brass industry created a new market for copper. In 1843, when there were three major mills in Waterbury and two more at Wolcottville and Derby, the industry consumed about 400 tons of copper. After substantial growth in the late 1840s and early 1850s, the Naugatuck Valley brass manufacturers were consuming about 2,000 tons of copper annually by 1855.[53]

The smelting industry that emerged in the mid-1840s was concentrated in a few east coast cities and used ores from South America and from the expanding domestic mines. The first burst of smelting was the result of several developments that made smelting economical: canal and railroad construction reduced the cost of anthracite coal at port cities; copper ore production in Chile and Cuba jumped dramatically; and in 1842 the British government increased duties on foreign ore smelted in Britain while at the same time the Swansea smelters sharply reduced the price they paid for both foreign and domestic ore. A second group of American smelting works began operating in the late 1840s, situated on or near the Great Lakes in order to use Lake Superior copper.[54]

The first new smelter to use foreign ore was the Revere Copper Company works, built in 1844 at Point Shirley, an isolated promontory in Boston harbor about four miles due east of the city. Point Shirley mainly smelted Chilean and Cuban ores, and in the late 1850s ores from Vermont, Tennessee, and Michigan. Domestic ores, however, never amounted to more than 10 percent of the total smelted. The Revere smelter closed in 1872, unable to survive the disappearance of foreign ores as a result of the duties the American government imposed in 1869. Point Shirley's peak production was about 5,000 tons of metallic copper from 25,000 tons of ore.[55]

Baltimore became a major center for copper smelting in the 1840s. Haslett McKim and Dr. David Keener incorporated the Baltimore & Cuba Smelting and Mining Company in 1845 and built a smelting works at Locust Point in Baltimore. The firm began smelting in 1846 with ore from Chile. When the early results were disappointing, the Baltimore & Cuba company closed the Locust Point plant in 1851. In dismantling the works in 1854, the owners found large quantities of metallic copper in the furnace bottoms and a mountain of discarded slag rich in copper. The sale of these materials to English smelters, combined with an influx of new capital, allowed the owners to reexamine all options, and in 1855 they built an even larger works at Locust Point.[56]

Keener organized the rival Baltimore Copper Smelting Company in 1850 and built a smelter at Canton, across the harbor from Locust Point. According to one report, the Canton smelter produced 3,000 tons of metallic copper and consumed about 15,000 tons of Cumberland coal in 1860 alone. The Baltimore & Cuba Company, under Clinton Levering's leadership, subsequently bought out the Baltimore Copper Smelting Company in 1864 and created the Baltimore Copper Company. The new firm operated both works until 1868, when the Baltimore and Ohio Railroad bought the Locust Point property to use as a terminal.[57]

Two other east coast copper smelters opened in the late 1840s, but information on them is skimpy. The Hendricks family built a smelter to treat sulphide ores at Bergen Point, New Jersey, at the site of their Soho (Belleville) works. The Bristol mine sent ore to Bergen Point starting in 1848, and the Vermont mines may have done the same. Bergen Point ceased smelting in 1866, one of several smelters unable to survive in the depressed economy following the Civil War. A smelter at East Haven (New Haven), Connecticut, treated ores from Chile, the Bristol mine, and the Copper Hills mine in Vermont. On the eve of the Civil War, it was still at work, owned by the New Haven Copper Company.[58]

Four additional smelters went into operation before the Civil War, all built to treat the native copper of the Lake Superior mines and specifically to smelt the large masses of nearly pure copper from those mines. The Pittsburgh and Boston Mining Company, the first firm to mine copper in Michigan's Keweenaw Peninsula, sent native copper from its Copper Harbor mine to the Point Shirley works in 1844–45 and in the following year shipped the first masses from its Cliff mine. Until 1849, virtually all the Michigan copper went to the Revere smelter in Boston and the Baltimore & Cuba works.[59]

American copper smelters were unable to handle the large masses of pure copper from the Lake Superior mines. The Baltimore smelter tried to treat the thousand-pound "nuggets" of Lake copper by opening one end of their reverberatory furnaces, winching the mass across the hearth, and then bricking up the opening. Early in 1847, Dr. Curtis G. Hussey, one of the Pittsburgh and Boston Mining Company directors, asked the Revere smelter's management to handle the masses, but they responded by threatening to charge him eighty dollars per ton, with no guarantee of the results. Hussey also tried to smelt Lake masses in a cannon-melting furnace at the Fort Pitt Foundry in Pittsburgh in 1847, but the loss of copper in the form of slag was excessive.[60]

In the spring of 1848, Hussey and other investors established C. G. Hussey & Company, with plans to construct a copper smelting works and rolling mill. To help build and then operate his Pittsburgh smelting works, Hussey hired two experienced Welsh smelters, William and Henry Johns, who had come to the United States in 1846 to work at the Locust Point works in Baltimore. According to his son, Henry Johns designed and built a reverberatory furnace in Pittsburgh with a removable top, which enabled large copper masses to be lowered directly into the hearth. Hussey built a smelting works and rolling mill on the Monongahela River about a mile north of Pittsburgh in 1849–50 and initially operated the works with about a hundred Welsh workers. His manufacturing complex, the Pittsburgh Copper & Brass Rolling Works, continued operating well into the twentieth century.[61]

Curtis Hussey's brother, J. G. Hussey, also operated a smelter in Cleveland from 1850 to 1867, but the two Hussey smelters treated the copper from only the Cliff and National mines, which the Husseys controlled. James Park Jr. and John McCurdy, two Pittsburgh industrialists, established the firm of Park, McCurdy & Company and in 1858 built a large smelting works and rolling mill in Pittsburgh using the Hussey reverberatory furnace design. Smelters built in Detroit in 1850 and in Hancock, Michigan, in 1860 also used Hussey's furnace design, as did David Keener when he built a furnace in 1860 at the Canton works in Baltimore.[62]

Starting in the 1850s, smelting moved even closer to the Michigan deposits. When Israel Coe of the Wolcottville Brass Company visited Detroit in 1849, he and John R. Grout of Detroit discussed the prospects of building a smelting works there. The following year, Grout visited Waterbury, where he and Coe persuaded the major brass manufacturers to invest in a Detroit smelter. They organized the Waterbury and Detroit Copper Company on 25 May 1850 and immediately built a smelter on the Detroit River about two miles south of the city. The smelter initially consisted of a single reverberatory furnace and a cupola furnace. By the onset of the Civil War, however, it had doubled in size and was producing nearly 3,000 tons of metal annually.[63]

Finally, in 1860, mine operators and other local capitalists built the Portage Lake Smelting Works in Hancock, Michigan, near the major mines in the district. The Portage Lake works began operating early in 1861, but then merged in 1867 with the Waterbury and Detroit Copper Company to form the Detroit and Lake Superior Copper Company. The new firm consolidated the two works under Grout's management, resulting in greatly improved efficiency. The 1860 Census of Manufactures returns for the seven largest smelting works show the scale of these operations (table 1.3).[64] The figures for Pittsburgh include two smelters and Hussey's rolling mill, so the capital investment is not extraordinary. The low capital investment for the Cleveland smelter is inexplicable. The returns included small smelters in New England but failed to report on smelting in Tennessee.

The consumption of copper worldwide expanded vigorously in the 1840s and 1850s, accompanied by shifts in the composition of demand. To be sure, most copper consumed in the United States

TABLE 1.3

Principal Smelting Works in the United States, 1860

City	Capital Invested	Number of Male Hands	Value of Product
Baltimore	$600,000	150	$1,300,000
Bergen Pt., N.J.	150,000	60	920,000
Boston	300,000	55	500,000
Cleveland	10,000	15	266,500
Detroit	100,000	40	1,500,000
Pittsburgh (2)	950,000	65	808,000

Source: U.S. Bureau of the Census, *Manufactures of the United States in 1860: Compiled from the Original Returns of the Eighth Census* (Washington, D.C.: USGPO, 1865), 43, 67, 221, 245, 272, 337, 447, 494, 598, 609.

TABLE 1.4

Gross Imports of Copper into the United States, Selected Years, 1840–1860, By Value

Yr/End.	Pigs, Bars and Old	Ore	Plates For Sheathing	Other Mfres	Total
30 Sept.					
1840	$1,100,664	—	$411,567	$141,084	$1,653,315
30 June					
1845	1,225,301	$48,807	738,936	111,211	2,124,255
1850	1,167,411	$195,382	715,614	339,323	2,417,730
1855	2,227,457	889,007	740,233	165,676	4,022,373
1859	925,488	1,346,501	156,891	121,168	2,550,048
1860	488,023	1,031,493	87,577	223,184	1,830,277

Source: New American State Papers: Commerce and Navigation (Wilmington, Del.: Scholarly Resources, 1973), 22:12–13; 29: 110–11, 118–22; 32:14–47, 178–81; 39:38–39, 174–77, 200–201; 43:16–17, 170–75, 180–81, 186–87, 196–97; 44: 176–79, 184–85, 200–201.

continued to go into sheets and sheathing, vats and vessels, and tubing, and into the manufacture of brass. One significant change involved ship sheathing. In 1832, George Frederick Muntz, a British inventor, patented a new sheathing metal, known as "yellow metal" or "Muntz Metal." The new alloy consisted of 60 percent copper and 40 percent zinc, making it considerably cheaper than standard copper sheathing. During the 1850s, Muntz Metal practically displaced copper for ship sheathing.[65]

Copper imports grew steadily from 1840 until the mid-1850s, but in the late 1850s, retained or net copper imports dropped sharply, reflecting several major shifts in the copper market. The Panic of 1857 triggered a general drop in demand, while domestic copper output soared in the 1850s. Because the rapid shifts in demand in the late 1850s were in part the result of uncertainty generated by increasing political tensions within the United States, the trade figures for both 1859 and 1860 appear in table 1.4.

During the decade of the 1840s, the value of brass imports approximated the value of domestic copper and brass exports, roughly canceling out each other. In 1855, brass imports were roughly $250,000, but copper and brass exports were nearly $750,000, so the gross import figures for 1855 in table 1.4 should be reduced by about $500,000. However, by 1859 and 1860, brass imports had dropped to about

TABLE 1.5

Estimated Domestic Production of Copper, Retained Imports, and Consumption, 1840–1860, in Tons

	Domestic Production	Retained Imports	Domestic Consumption
1840	0	4,000	4,000
1845	112	5,000	5,112
1850	728	6,000	6,728
1855	3,904	8,000	11,904
1859	7,056	4,500	11,556
1860	8,064	2,500	10,564

Source: William B. Gates, *Michigan Copper and Boston Dollars: An Economic History of the Michigan Copper Mining Industry* (Cambridge, Mass.: Harvard University Press, 1951), 197, 201.

Note: Value figures for imports were converted into tonnage by dividing the dollar value of imports by an average of the prices for old copper and copper sheathing given in Anne Bezanson, *Wholesale Prices in Philadelphia, 1784–1861* (Philadelphia: University of Pennsylvania Press, 1937), 2:53–55. I have revised Gates's national output estimate for 1855 upward by about 500 tons. Gates combines Michigan production for 1855 (2,904 tons) with a figure of only 456 tons for the rest of the country. The Ducktown district alone produced at least 800 tons in 1855, so I have estimated the non-Michigan copper output at 1,000 tons for 1855.

$200,000, while exports of copper and brass products were about $1 million. Retained imports of copper, which were valued at about $3.5 million in 1855, fell to about $1.75 million in 1859 and to $1 million the following year.[66]

The American copper mining and smelting industries enjoyed little tariff protection before the Civil War. There were no duties on copper ore until 1861, and then only a modest duty of 5 percent ad valorem, while regulus, or matte copper, remained duty-free until 1869. The Tariff of 1846 placed a 5 percent duty on ingot and scrap copper but a more protective duty of 30 percent on manufactures of copper excluding sheathing. The Tariff of 1857 returned ingot and scrap copper to the free list and reduced the duty on copper manufactures to 24 percent, a medium level of protection.[67]

Total consumption of imported copper was about 4,000 tons in 1840 and then climbed to 6,000 tons in 1850, including around 500 tons derived from foreign ore. In 1855, imported copper retained for domestic use, including about 2,000 tons from imported ore, amounted to about 8,000 tons. Table 1.5 shows the dramatic impact of the Michigan copper district in supplying the American market in the years just before the Civil War. On the eve of the Civil War, retained imports fell to about 2,500 tons, reflecting the rise of the domestic copper industry and growing American self-sufficiency in the red metal.[68]

CHAPTER 2

Michigan Copper through the Civil War

The collection of minerals I have brought from the regions of Lake Superior have turned the heads of most of those persons who have examined them, but it is not so with myself, for I know full well the many difficulties and embarrassments which will surround the development of the resources of this district of [the] country.—Douglass Houghton, State Geologist of Michigan, to Augustus Porter, member of Congress from Detroit, 26 December 1840

The first Europeans to visit North America found that Native Americans in the eastern United States had copper ornaments, weapons, tools, and nuggets. More recently, burial mounds, graves, and village sites in the Mississippi Valley and points east have yielded a great number and variety of copper artifacts. Much of this copper came from the south shore of Lake Superior.

The archaeological evidence suggests that Native Americans mined copper in this region continuously from about 3,000 B.C. through the sixteenth century and traded it throughout the Mississippi Valley and the southeastern United States. Once superior European goods—including iron knives, hatchets, and other tools, along with brass kettles—were available to the Native Americans of the region, they acquired copper mainly for ornamental and ceremonial uses.[1]

Explorers who came to the Keweenaw Peninsula in the 1840s found hundreds of shallow pits dug by Native Americans centuries earlier. Charles Whittlesey summarized these discoveries in a report

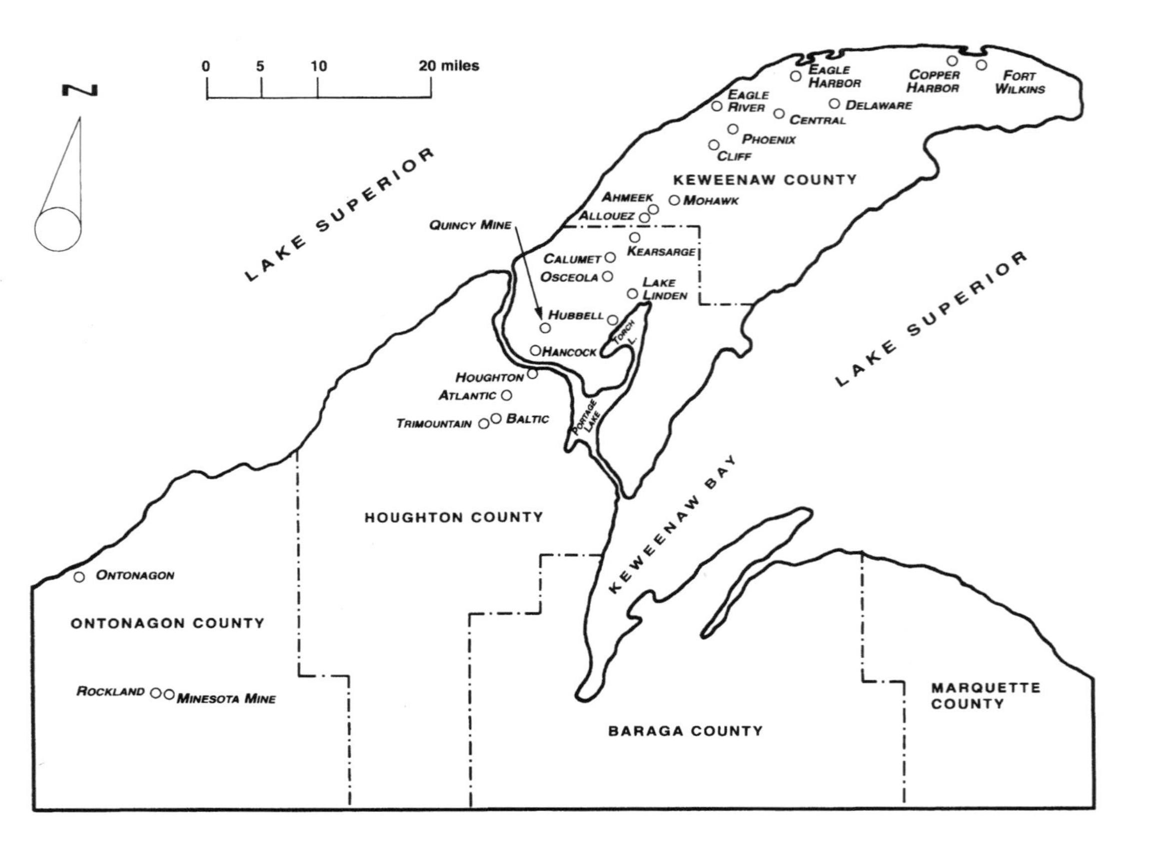

The Michigan copper range. (Map by Mike Brooks)

submitted to the Smithsonian Institution in 1856 but not published until seven years later. He described hundreds of pits that sometimes were twenty to thirty feet deep and usually held stone hammerheads, or mauls, often found in the hundreds in a single pit, along with wooden shovels, copper chisels, and copper spearheads. Evidence of fires suggested that these miners may have heated the stone that encased the copper and then fractured the rock by throwing water on it.[2]

When Europeans arrived on the Great Lakes in the early seventeenth century, Native Americans were working these shallow pits only haphazardly, if at all. Most of the available copper consisted of fragments cut from "float copper" found on the surface and nuggets retrieved from river bottoms and shorelines. Still, the real and imagined copper deposits on Lake Superior attracted explorers and motivated federal and Michigan authorities to send geological expeditions into the region.[3]

The first, led by Michigan territorial governor Lewis Cass, with Henry R. Schoolcraft serving as mineralogist, explored the south shore of Lake Superior from Sault Ste. Marie to the site of present-day Duluth in the summer of 1820. The Cass / Schoolcraft expedition increased public awareness of the existence of copper in the Lake Superior region. Schoolcraft later submitted a lengthy report to Secretary of War John C. Calhoun in which he reported the discovery of native copper in the area. Schoolcraft published this document, with Calhoun's permission, in the *American Journal of Science and the Arts*. He also wrote a book that summarized the major findings of the expedition and included a daily journal of the trip.[4]

In 1831 and 1832, Schoolcraft led two more expeditions into Lake Superior and the upper Mississippi Valley, under the pretense of diffusing Native American hostility. Schoolcraft selected Dr. Douglass Houghton to serve as surgeon and naturalist for both expeditions. During the expeditions of 1831 and 1832, Houghton visited the Ontonagon Boulder, a two-ton piece of "float copper" found on the Ontonagon River forty miles from Lake Superior. He also examined the nearby lands for other signs of copper but found none. In his official reports, Houghton remained cautious in assessing the mineral resources of Lake Superior, reminding others that the many masses of "float copper" found around the Keweenaw Peninsula had not led to the discovery of underground deposits.[5]

After Henry Schoolcraft negotiated the Treaty of Fond du Lac with the Chippewa in 1826, the United States government could explore for minerals on Chippewa lands, but private parties still could not. In achieving statehood in 1837, Michigan reluctantly accepted the western Upper Peninsula as compensation for ceding the disputed "Toledo Strip" to Ohio. The new state legislature established a state geological survey and appointed Douglass Houghton to the post of state geologist. Houghton surveyed Michigan's Lower Peninsula in 1837–39 and in the summer of 1840 began exploring the south shore of Lake Superior.[6]

In his fourth annual report on the geological survey, issued in February 1841, Houghton cautiously suggested that the Keweenaw Peninsula contained significant copper deposits. He shared his findings with his fellow geologists at the second meeting of the Association of American Geologists and Naturalists, held in Philadelphia in April 1841. A summary of those remarks appeared in the *American Journal of Science*, disseminating the information to an even wider audience. Houghton, however, was more enthusiastic about the region's prospects in private communications in December 1840 to Augustus Porter, the U.S. representative from Detroit. Houghton reported:

> The chief of the important minerals that occur in this district are the different ores of copper; which are found in veins traversing the rocks of the mineral region; and these veins are of a width that will not suffer in comparison with veins of a similar character in any known mining region.

Excerpts from his letter to Porter appeared in the *Detroit Daily Advertiser* in May 1841, and helped launch the "copper rush" to Lake Superior.[7]

Michigan Copper Fever

Houghton's work added to the pressure on Congress to buy the mineral lands from the Chippewa. In the Treaty of La Pointe, which took effect on 23 March 1843, the Chippewa sold to the United States all their lands between present-day Marquette and Duluth, including the entire copper district. Later in 1843, Julius Eldred, a Detroit hardware merchant, moved the Ontonagon Boulder to the lakeshore and placed it on a ship bound for Detroit. The United States government claimed ownership of the famous nugget, confiscated the prize, and moved it to Washington. The controversy received considerable attention in the eastern press and generated even more public interest in the copper district.[8]

Federal land policy encouraged a rush of prospectors and speculators to the district, producing the "Michigan copper fever" that began in the summer of 1843 and extended through 1846. Using the precedents established for reserving and leasing lead lands, Secretary of War James Porter established a United States Mineral Land Agency at Copper Harbor at the tip of the Keweenaw Peninsula in April 1843, with General Walter Cunningham serving as the first agent. The copper rush began slowly. Only about twenty prospectors reached Copper Harbor in 1843, and most of them spent the winter of 1843–44 in the district. An individual could obtain a permit that gave him the right to explore a tract not exceeding nine square miles for one year and to apply for a three-year renewable lease on the tract, paying as rent 3 percent of the value of any metal extracted. The leaseholder had to post a $20,000 bond. In March 1845, Secretary of War William L. Marcy reduced the maximum size of a tract to one square mile.[9]

Through 1844, while the flow of prospectors and speculators was still small, the War Department remained in control. The reports that John Stockton, the superintendent of U.S. mineral lands on Lake Superior, wrote to the War Department in February 1845 focused on the geological and land surveys conducted by his assistants. A year later, Marcy reported to Congress on the state of the mineral lands in Michigan and included a revealing summary of mining activity through 1 September 1845. During the summer of 1845, an average of 157 men worked in mining and exploration. They sank a total of only eleven shafts, with a combined depth of 300 feet, at four locations. The total amount of copper rock removed from the earth was a mere 123 tons, with only 17 tons shipped out of the district. Equally distressing was the cost of administering the leasing system through April 1846, estimated at $32,800, which was split evenly between salaries and other costs, including buildings, equipment, and supplies.[10]

At the urging of Congress, Marcy suspended the issuance of new leases on 6 May 1846. Up to that point, prospectors and speculators had acquired more than a thousand permits to explore and had leased 58 tracts of nine square miles and 317 tracts of one square mile. Holding permits and leases, however, did not guarantee that serious exploration work, much less mining, would follow. Total royalties collected since 1843 amounted to $192, reflecting the lack of significant mining activity.[11]

Investors could speculate in copper mines by buying shares in the many companies that sprouted like mushrooms on a warm summer morning. Eastern investors, who often had no reliable information, were participating in a "subterranean lottery" when they bought shares in these early mining ventures. Newspapers routinely reported preposterous stories about discoveries. *Niles' National Register,* for example, confidently reported that a vein found near Fort Wilkins at the tip of the Keweenaw Peninsula would easily produce $6 million per year in metal. It proved worthless.[12]

Many early mining companies were merely paper ventures launched to bilk naive investors. By the end of 1845, about forty-five "companies" existed, but only thirteen of them were doing any work at their locations, and only five ever produced any copper. The 65 companies that existed in early 1846 grew to 86 in July and to 102 in October. The speculative fever in stocks peaked in 1846, when a handful of manipulators made large fortunes. One visitor estimated that three-quarters of the people managing the development of new mines in 1846 were "dishonest speculators and inexperienced adventurers."[13]

The speculative bubble finally burst during the summer of 1846. A War Department directive issued in May 1846 invalidated most of the permits and leases previously awarded to speculators. Uncertainty concerning ownership and pessimism over the future of the copper mines produced a sharp drop in copper mine share prices in August 1846, which marked the end of the first Michigan copper fever. In 1847, Congress gave control over land sales to the Treasury Department, which then sold mineral lands in quarter sections at a minimum price of $2.50 per acre. Congress reduced the minimum price to $1.25 per acre in September 1850, further encouraging long-term investment in the district.[14]

As more detailed geological and economic information about the Michigan copper deposits developed, potential investors were better informed about condition in the Lake Superior district. Two valuable handbooks appeared in early 1846: *A True Description of the Lake Superior Country,* by John R. St. John, and a collection of materials compiled by Jacob Houghton Jr. and T. W. Bristol, *Reports on the Mineral Region of Lake Superior.* Both provided detailed descriptions of the geography and geology of the copper district, reports on the leasing system, summaries of individual mining companies, and glossaries of geological and mining terms.[15]

Horace Greeley, a shareholder and director of the Pennsylvania Mining Company since 1845, visited the Keweenaw Peninsula in 1847 and 1848 and reported extensively in the *New York Weekly Tribune.* He pointed out the enormous expense and risk involved in opening a new mine, warning investors that they needed at least $50,000 to develop a new property. Other sources of information included the first regular newspaper published in the region, the *Lake Superior News and Miners' Journal,* begun in July 1846 at Copper Harbor.[16]

The initial development of Michigan's copper district did not produce impressive results. Total production of metallic copper for 1846 was about thirty tons, with a single mine (the Cliff) accounting for two-thirds of the total. However, the "copper rush" of 1841–46 made future explorations easier and later investments less risky. Speculation brought people to an isolated, rugged wilderness with no other attractions. By 1850 the district could boast of 1,097 permanent white residents. Once settlements like Copper Harbor and Eagle River could serve as base camps, exploration was less formidable.[17]

After the Michigan copper fever had subsided, investment proceeded in a more orderly and productive fashion. A handful of eastern capitalists made substantial investments in more than a dozen mines, and the spectacular success of two, the Cliff and the "Minesota," encouraged even greater involvement. The results were impressive, especially when compared with the previous period of speculation. By 1855 the Michigan district had generated about 2,900 tons of metal, with thirteen mines producing at least thirty tons of copper each and another seven with at least some output.[18]

The Pittsburgh and Boston Mining Company's success in the late 1840s illustrates the risks and the rewards of investing in copper mines in this district. The firm began in 1843 as a partnership controlled by Pittsburgh capitalists and became a joint-stock company in May 1844. The company developed two locations near Copper Harbor in 1844–45, when it spent $28,000 and realized about $3,000 in revenues before quitting that location in 1846. At the Cliff mine near Eagle River, opened in 1846, the initial explorations cost $66,128 but yielded only $8,870 from sales of copper. The company directors raised additional funds by assessing the outstanding shares a total of $110,000, but when the stockholders balked in early 1847, one director, Charles Avery, advanced $80,000 so that the company could continue the development work.[19]

The Cliff mine required even more time and capital before it began to pay its investors. Production climbed to 205 tons of metal in 1847, but the resulting revenue of $71,000 went back into further development work. The following year, output reached 498 tons and revenues were $166,000, which the company again reinvested in the mine. The Cliff did not reward the stockholders until May 1849, when it paid dividends of $60,000, the first from a Lake Superior copper mine. During the firm's first six years of operation, the stockholders paid assessments totaling $110,000 and allowed all the revenues, about $250,000, to be plowed back into the venture before they realized any returns. After that, the Cliff mine paid regular returns from 1849 through 1867, except for 1860, for a total of $2.5 million in dividends through 1879. Although the Cliff mine was rich in copper and easy to open, development took four years and required large expenditures of capital.[20]

The second successful producer, the Minesota mine, was so named as result of a spelling error when a clerk filled out the required incorporation documents for the State of Michigan. Discovered in 1847, the Minesota began producing the following year, but its stockholders paid assessments of $366,000 before the mine paid its first dividend in 1852. It enjoyed a few productive years and paid dividends of $1.76 million by 1864. The Minesota mine earned the contradictory nicknames "Mother of Mines" and "Father of Failures," reflecting the influence it had on investors in the rest of the copper region. The Cliff and Minesota mines, however, were exceptional properties blessed with massive copper deposits—in 1855 they accounted for more than two-thirds of Michigan's copper output.[21]

The history of the less successful mines also bears repeating. The same Pittsburgh investors who developed the Cliff mine founded the Northwestern Mining Company in 1845 and immediately began explorations but did not ship any copper until 1852. The company suspended operations in 1857 but resumed in 1863, only to stop permanently at the end of 1864. By then the firm had called in $228,000 in assessments on a capital stock nominally worth $300,000. The Northwestern also plowed back $80,000 realized from the sale of copper, making a total investment of $308,000, but paid no dividends. Other prominent mining ventures that failed after substantial investments included the Copper Falls Mining Company and the Northwest Mining Company.[22] The success of the Cliff and Minesota mines gave investors hope that their stock would eventually earn them a fortune. Gates has argued that the prospect of large profits, however remote, was sufficient to generate a large, steady flow of capital into the district:

> The gambling element is the key to understanding (the influx of capital), and it is even a mistake to measure pecuniary success of individual stockholders chiefly in terms of dividends. The game was so uncertain, and there was such a substantial lapse of time between the day a mine opened and the day it was finally proven a success or failure that there were plenty of opportunities to make a fortune by speculating in the securities of even those companies that ultimately ended completely bankrupt.[23]

Gates to the contrary notwithstanding, eastern capitalists did not invest blindly in unknown properties. By the early 1850s, investors had obtained access to detailed information on the copper district. The geological reports of Charles T. Jackson, John W. Foster, and Josiah D. Whitney in 1849 and 1850 provided detailed information on each mine location. In a more popular vein, Robert Clarke's reports on the copper region in *Harpers* in 1853 and Whitney's *Metallic Wealth of the United States,* published in 1854, further informed the public. National commercial publications such as *Niles' National Register* and *Hunt's Merchants' Magazine* devoted space to Michigan copper. With the appearance of the *Mining Magazine* in July 1853, Lake Superior copper mines received extensive regular coverage from a well-informed correspondent in Houghton, in the heart of the copper district.[24]

Through the stock assessment system, investors could proceed cautiously and minimize their risks. All the copper mining companies issued stock requiring a small amount of paid-in capital compared with the face value, usually less than 10 percent. As the firm developed the property and needed additional capital, the directors could levy further assessments, usually calling for payments of 5 or 10 percent of the face value of the shares. Each call had to be justified, and the individual stockholder who wanted to avoid further involvement could sell his stock or simply refuse to pay the assessment and forfeit the shares.[25]

The stock assessment system also allowed large investors to buy small stakes in dozens of mines and then cull out the weaker properties later. A few substantial eastern capitalists became large stockholders and directors of many mining firms: Horatio Bigelow (Boston) was the treasurer of fourteen mines in 1854; the tandem of Thomas M. Howe and Curtis G. Hussey, both from Pittsburgh, were directors of at least eight mines in the late 1860s; and in 1865 the partnership of T. Henry Perkins (Boston) and Thomas F. Mason (New York) served on the boards of seven mining companies that accounted for nearly 40 percent of Michigan's copper output.[26]

A detailed history of investment in one struggling copper company shows how cautious and hard-nosed eastern capitalists could be when investing in Michigan copper mines. The Quincy Mining Company, founded in November 1846 in Marshall, Michigan, about a hundred miles west of Detroit, incorporated on 30 March 1848 with capital stock of 4,000 shares valued at $50 per share. Because the major stockholders lived in Detroit, the company held its meetings there beginning in 1847. After committing considerable capital to the venture and seeing no return in sight, the Detroit investors sold the property to Philadelphia capitalists, who invested heavily in the mine between 1851 and 1856, again without success. Another set of Detroit investors held the property from 1856 to 1858, but a new combination of capitalists from New York finally made the mine pay. The Quincy Mining Company paid its first dividend in 1862 after investors had sunk nearly $500,000 into the venture. The firm then paid dividends so regularly between 1862 and 1920 that it became known as "Old Reliable" in mining circles.[27]

It is a commonplace that Boston-based capitalists developed the Michigan copper district. Professional and amateur historians alike have focused on the Bostonians. In his classic study, *Michigan Copper and Boston Dollars: An Economic History of the Michigan Copper Mining Industry*, William B. Gates Jr. argued that by the early 1850s the Boston Stock Exchange was the principal source of capital for the Lake Superior copper mines. Gates, however, could not explain this dominance. More recently, however, Collamer Abbott reexamined the development of Boston involvement in Michigan mines and argued that other financial centers provided considerable capital as well.[28]

During the initial wave of investment, which extended through 1855, Boston stood at best on a par with other sources of capital. The location of company offices is a good indicator of the major source of capital, because the largest stockholders usually placed the company office in their home city. Of sixty-five companies working in late 1845, sixteen were in Detroit and fourteen in New York. Thirteen companies had offices in Michigan cities other than Detroit, while Boston could claim the same number, or one-fifth of the total, hardly a position of dominance. In July 1846, when eighty-six companies were in business, Boston investors controlled twenty-five, less than one-third of the total.[29]

The results were similar a decade later, after most of the weaker companies had disappeared. An 1855 listing of thirty-five mining companies showed eight with their headquarters in New York, eight in Boston, seven in Pittsburgh, and seven in Michigan cities, including Detroit. The distribution of copper production is even more striking. Of total Michigan copper output through 1854, Pittsburgh-based companies accounted for 58 percent, New York firms 34 percent, and

Boston-based companies a mere 5 percent. The Cliff mine, with capital from Pittsburgh, and the Minesota, a New York company, dominated the district through the mid-1850s.[30]

The Rise of the Lake Superior District

A more orderly investment market emerged after 1855, when the Boston Stock Exchange first listed Michigan mining stocks. Initially only eight companies had their shares regularly quoted, but by 1860, the number had increased to twenty-one. Detailed information about individual properties was readily available by the late 1850s. Two copper district newspapers began during these years—the *Lake Superior Miner,* founded in Ontonagon in 1855, focused on the southern end of the mineral range, while the *Portage Lake Mining Gazette,* established in 1858 at Houghton, covered the central and northern part of the district. By the late 1850s, R. G. Dun & Company was producing credit ratings for many mining firms.[31]

Most mining companies wanted to create the impression of solidity and success, so they issued elaborate annual reports to their stockholders. These often included a detailed breakdown of expenditures, costs, and revenues, so a discerning reader could learn much about the company's real condition and future prospects. The Pewabic Mining Company, for example, started mining in 1855 and issued its first annual report in 1858, although it did not pay dividends for another four years. The number of Michigan copper companies that produced printed annual reports peaked at twenty-five in 1853, then fell to eleven for the rest of the decade.[32]

On the eve of the Civil War, the Michigan copper district already had several major features it would retain for the rest of the nineteenth century: domination by a few firms and a heavy concentration of production in two or three districts. Twenty-two mines produced copper in 1860, but the ten largest accounted for 96 percent of the total. The Minesota mine remained the largest producer, while the Cliff had fallen to fourth place. Three neighboring Ontonagon County mines (Minesota, National, and Rockland) together produced 37 percent of the total for the district, while the Portage Lake mines (Quincy, Pewabic, Franklin, and Isle Royale) accounted for a similar share.[33]

The Civil War brought a new speculative fever. The demand for war goods, combined with price inflation, created an expansionary climate for the copper industry. Prices jumped from 19 cents a pound in 1861 to 46 cents in 1864. The number of dividend-paying companies increased from three to eight between 1861 and 1864, while the total dividends paid out annually jumped from $120,000 to $1.2 million over the same span. Dozens of new companies also emerged. The number of copper stocks quoted on the Boston Stock Exchange held steady at twenty-one in 1859–1861 and then increased quickly to fifty-two in 1864 and 1865. Similarly, the number of companies that issued printed annual reports jumped from fourteen in 1860 to sixty in 1865, the height of the wartime fever. Objective observers noted that "a liberal proportion of the new companies have been organized for the benefit of the founders rather than for that of the stockholders."[34]

The development of new mines and the speculation in stocks had negative effects on the district as a whole. While some mines, like the Minesota, experienced falling output because of depleted deposits, others had to reduce production because of labor shortages. The demand for miners to open new mines exacerbated the general manpower crisis caused by enlistments and the draft. Michigan's 1861 production of 7,518 tons was the wartime peak. Output fell to a low of 6,245 tons in 1864 before recovering to 7,197 tons during the last year of the war. The ten top producers in 1865 accounted for three-quarters of total output but included only one mine not already operating in 1860.[35]

Although the Boston Stock Exchange had functioned as the major formal market for copper shares for a decade, Boston investors still did not dominate the Michigan mining scene at the end of the war. Boston-based companies held one-third of the paid-in capital of all the Michigan copper companies in 1865 and produced about one-third of the annual reports but still accounted for only 23 percent of the district's output. At the end of the war, companies based in Pittsburgh and New York still produced two-thirds of the copper, but the shift toward Boston domination was beginning. An article about the decline of copper stock prices in 1864 blamed the drop on a general withdrawal from copper shares by New York and Philadelphia investors, who had suddenly shifted their funds into petroleum issues.[36]

The development of the Michigan copper deposits had begun as wild speculation, a form of gambling in "underground lotteries," but had evolved to a more orderly, rational commitment of venture capital by investors in eastern cities a thousand miles from the copper lands of Lake Superior. The locus of investment had shifted from Michigan to more highly developed cities in the Midwest, such as Cleveland and Pittsburgh, and ultimately to Philadelphia, New York, and Boston, where investors were wealthier and more sophisticated than their midwestern counterparts.

Scholars and popular historians alike have exaggerated the importance of unbridled speculation in the emergence of the Michigan copper region. One local newspaper claimed, without evidence, that speculators established more than 300 companies between 1841 and 1865. Investors did launch at least 94 joint-stock mining companies through 1865, with a combined nominal capital of $48 million but paid-in capital of only $12.4 million. The companies, however, also realized a total of $39.4 million from sales though 1865 and perhaps reinvested about a tenth of that amount, making the total investment around $17 million.[37]

Only 8 of the 94 companies paid dividends, amounting to $5.6 million in total, while 52 of them produced no copper through the end of the Civil War. Another 13 mined less than thirty tons each. In short, two-thirds of the copper ventures, with half of the direct investment, failed to produce significant amounts of metal. In aggregate terms, however, the results were impressive. Copper production increased from a mere 51 tons in 1848 to 640 tons in 1850, then jumped to 2,900 tons in 1855 before doubling to 6,000 tons in 1860. After peaking at 7,518 tons in 1861, output fell to 6,245 tons in 1864 and then recovered to 7,179 tons in 1865. Michigan accounted for three-quarters of U.S. copper production between 1845 and 1865, a period of rapid growth in the industry.[38]

Machines and Men

Copper occurred in Michigan in its native state, as pure metallic copper, unalloyed with other elements. The native copper appeared as three distinct types of deposits. The first, fissure deposits, contained large masses of pure copper that formed in faults in the earth's geological formations. The masses ranged in size from twenty or thirty pounds to about 500 tons, but single masses larger than 100 tons were rare. The second type of copper deposit resulted when water deposited copper in the almond-shaped vesicles (voids) left in rock formed by lava flows emanating from fissures found throughout the Great Lakes Basin. These are known as amygdaloid deposits, from the Greek word for almond (*amygdule*). Finally, the third type of deposit, conglomerate copper, resulted when copper filled the existing spaces with a mixture of sand, pebbles, and stones, binding them together and creating a ribbon of copper.[39]

The categorization of copper mines as mass, amygdaloid, and conglomerate is an oversimplification. Nearly all the mines had some mass copper, and most fissure mines had amygdaloid or conglomerate deposits as well. The first generation of successful mines, including the Cliff, Copper Falls, Central, Minesota, National, and Rockland, exploited fissure veins. In 1865, however, the fissure mines accounted for only one-fourth of copper output, and the amygdaloid mines of the Portage Lake district had already surpassed them. The mass mines exhausted their deposits in the 1870s, and over the working life of the Lake Superior district, amygdaloid and conglomerate mines accounted for 97 percent of production.[40]

Most mines produced three distinct products—mass copper, which was nearly 100 percent metal; kiln copper, or "barrel work," which were fist-sized pieces of copper largely cleaned of adhering rock and packed in barrels for shipment to the smelter; and stamp rock, or "stamp work," which contained about 2–6 percent copper before treatment. The mix of output at two Portage Lake amygdaloid mines during the Civil War, the Franklin and the Pewabic, was typical. Stamp work accounted for between 55 and 70 percent of output, kiln copper between 15 and 30 percent, and mass copper between 5 and 15 percent.[41]

Copper mining technologies came to Lake Superior in the heads and hands of the Cornish miners who migrated there. The reproduction of Cornish mining technology in the Michigan mines is convincing testimony about the importance of human agents in the transfer of technology in the nineteenth century. Descriptions of mining techniques used in Cornwall, whether taken from the 1770s or the 1850s, could be repeated with few modifications for the Michigan copper district through the Civil War.[42]

The technology employed to mine mass copper and copper-bearing rock remained primitive, labor-intensive, and largely unmechanized well into the 1850s. Miners sank vertical shafts on or near the copper lodes, drove tunnels called drifts along the lodes, and excavated tunnels known as crosscuts, which ran perpendicular to the drifts and aided in exploring any "side branches" extending from the main lode. They excavated copper-bearing rock by creating openings known as stopes in areas above and below the drifts. By striking handheld chisels with sledgehammers, miners created shots (holes some four feet deep) and filled the holes with black powder, which they exploded to free the rock. A second group of underground workers, the trammers, moved the rock to the shaft by wheelbarrow or on wheeled tramcars on iron tracks. At the shaft, they dumped the rock into a kibble (bucket) or a wheeled rock carrier known as a skip, and a windless or steam hoist in turn pulled the kibble or skip to the surface. Mechanized technology came into play only at this stage of mining operations.[43]

On the surface, men sorted the rock and discarded the "poor rock" containing no copper. The remaining rock went to a "kiln house," where heating in large timber fires broke up the larger pieces and separated the rock surrounding the copper masses from the metal. Men also broke the rock with sledges, throwing water on the hot rocks to aid the process, much as Native Americans had done centuries earlier. After gangs of men "burned and dressed" the copper, they sent masses of various sizes directly to the smelter, while the rest of the rock went to the stamp mill. There, mechanical "stamps," which operated like large mortars and pestles, crushed the rock into progressively smaller sizes, ultimately liberating the specks or ribbons of copper from the surrounding rock. Enormous quantities of water moved the materials within the stamp mill and carried away the lighter waste rock, which comprised between 96 and 98 percent of the material treated. The stamp mill produced "mineral," a concentrate containing 60 to 80 percent copper, which the mining company then shipped to a smelter. There, the process for producing pure metallic copper in reverberatory furnaces was simple, because Lake copper was free of other elements.[44]

Much of the technology in use in the Michigan copper district in the 1840s underwent significant mechanization by the end of the Civil War. Powerful steam hoisting engines pulled wheeled skips on tracks with wire rope, replacing hand-cranked windlasses. Steam-driven pumps replaced bailing as the means of removing water from the mines. Similarly, a steam-driven "man-engine," by which men moved in and out of the mine by riding rods driven by the engine, replaced ladders. By the Civil War, most mines used steam railroads or gravity tramroads to deliver copper rock to the stamp mills, which were highly mechanized processing plants.[45]

The labor force at the Michigan copper mines consisted almost entirely of adult males. In Cornwall, women routinely made up one-fifth of the mine workforce, while children under age sixteen accounted for another fifth. Women and young children worked on the surface, mainly in sorting, dressing, and stamping ore. Boys normally began working at age eight or nine, while girls started at nine or ten. Boys rarely went underground until age twelve but instead worked on the surface at the physically demanding tasks of wheeling rock or operating hand-driven bellows and pumps.[46] Lake Superior copper mines, however, did not employ women or girls in any capacity throughout the nineteenth century, while the few boys working underground or in the stamp mills were usually fourteen or fifteen when they began work. The 1870 Census delineated a total labor force for the copper district of 4,188, which included only twelve boys working aboveground and none working underground. The 1880 Census showed a total of 155 boys employed, roughly 3 percent of the district labor force of 5,004. Of these boys, 135 worked aboveground. The families of Cornish miners who migrated to Lake Superior thus enjoyed a significant improvement in the quality of life.[47]

Michigan miners in the nineteenth century labored under a contract system that was a greatly modified version of the "tutwork" system of labor used in Cornwall. There, copper and tin mines used two systems for organizing labor: tribute and tutwork. The tribute system, normally used for "paying ground" (areas of mine containing metal) involved groups of tributors called "pares" who worked a section of the mine, known as a "pitch," for a share of the revenues the owners realized from the copper sold. An auction, usually held every two months, awarded the pitches and established the terms of the contracts. Tutwork, however, involved groups of miners contracting to excavate ground at a fixed rate per cubic or linear fathom and normally applied to shaft-sinking and drifting through barren ground, where the tribute system could not be used.[48]

Michigan mines occasionally used the tribute system, often when the mining company was capital-poor, either at the very beginning of operations or at the very end. Most mining companies avoided tributing because of the difficulty of monitoring the tributors. Quincy, for example, used tributors between 1855 and 1858, in part as a strategy for exploring the property at little direct expense. The *Lake Superior Miner* reported five mines working on tribute in 1857, with all the parties pleased with the arrangement. The managers of the Cliff mine briefly considered working the property at tribute in 1871 and finally did so in 1879, but only as a last resort.[49]

By the mid-1860s, the contract system of mining was firmly in place, and this system remained largely unchanged into the twentieth century. The mining captain would let contracts, usually at the beginning of each month, to an individual contractor, who would then select a team of men, often relatives, to work with him. The team, called a "pare" in Cornwall and a "pair" in Michigan, normally had an even number of partners, because half worked each shift. The minimum size of a "pair" or a "party" was four men, the number needed for "single-jacking" (driving holes with one man holding the drill and one striking it with the sledgehammer), but six-man parties, needed for "double-jacking," were the norm.[50]

When the mining captain let contracts, he would establish a fixed pay rate for each cubic fathom of ground excavated, with the rates varying according to the type of work (shaft-sinking, drifting, crosscutting, or stoping) to be done and the quality (hardness) of the rock. The captain awarded contracts through an informal process of negotiation, so the rates in part reflected the supply of available contractors. However, there is no evidence of the competitive auction system found in Cornwall. At the end of the contract period, normally a month, the captain would measure the ground excavated. The mine clerk would deduct the cost of the supplies the party had used—including drill steel, candles, powder, and fuse—from their earnings before paying them. The miners usually divided their net earnings equally among the members of the party.[51]

The labor system used in the Michigan mines was a barely recognizable remnant of the Cornish mine labor system. It was not a piece-rate or incentive system despite its appearance. The mining captain set the contract rates to achieve a predetermined average monthly wage for the miners and adjusted rates accordingly. Mine managers were not reluctant to challenge traditional practices. In April 1855, Franklin Hopkins, the resident agent for the Northwest Mining Company, moved the "settlement day" for contracts from the last day of the month, the traditional practice in Cornwall, to the Friday afternoon or Saturday morning closest to the end of the month. The miners demanded the following Monday off as an "idle day," also a Cornish tradition, but Hopkins refused to give in, withstood a week-long strike, and imposed the new rules. The contract system used in the Michigan mines was a social fiction created and maintained to allow the Cornish miners to believe that they were not common wage laborers.[52]

The proportion of employees working on contracts and on "company account" (that is, for wages) varied greatly over time, as the mix of work and company policies changed. All the men employed by Quincy in 1851 worked on company account because most of the work done was general surface labor, traditionally not covered by contracts. However, over the next thirty-eight months, through February 1855, one-third of the men normally worked on a contract basis and two-thirds on company account.[53]

Labor practices remained flexible into the 1860s. In 1859 and 1860, when Quincy used the contract system for many tasks besides mining, two-thirds of the men worked on a contract basis. The company awarded contracts for operating windlasses and hoists, "breaking and assorting," dressing copper, wheeling rock underground, and for "running rock into the stamp mill." However, after 1861, when the stamp mill workforce, all wage workers, made up one-sixth of the entire labor force, contract workers never exceeded one-third of the Quincy workforce.[54]

The Michigan copper district had few strikes or other serious labor unrest until the Civil War. Labor peace was a reflection of the district's growth and general prosperity, which produced high wages and steady employment. To be sure, not all jobs were equal in terms of pay or difficulty. The monthly pay of the Quincy Mining Company's 620 employees in June 1865 varied from $350 earned by the superintendent to $15 paid to a boy employed as a washer in the stamp mill. In the middle ranks, lower-level officials like the boss machinist earned between $70 and $100 per month; skilled workers such as miners, blacksmiths, and carpenters earned between $50 and $60; unskilled or "common" laborers earned from $35 to $45; and only apprentices and other boys made less than $35 per month. Since wages remained high during the Civil War, labor peace was the rule.[55]

CHAPTER 3

Michigan Copper in Prosperity and Decline

Among copper mining companies the Calumet and Hecla stands alone. . . . The mine so greatly exceeds all others in extent and richness that there is none to be compared with it in product or profit. If any comparison is instituted it must be borne in mind that the Calumet and Hecla lode is probably by far the richest vein ever known in the annals of copper mining.—*Annual Report of the Commissioner of Mineral Statistics of the State of Michigan for 1880.*

During the quarter century that followed the Civil War, Michigan's copper output grew explosively, often exceeding the growth in domestic demand and generating copper prices too low to allow most producers to earn profits. The development of the rich Calumet and Hecla mines in the late 1860s brought an enormous growth of copper production at costs well below those of other producers. The Lake producers tried to gain control of the U.S. market through increased tariff protection and a variety of pooling arrangements, but the efforts to control the American and eventually the world copper markets were ultimately unsuccessful. By the late 1880s, the emergence of substantial low-cost mines in Montana and Arizona heralded the end of Michigan's dominant position.

Michigan's copper industry continued to grow from the early 1890s through World War I, but by the late 1910s the bleak future of the district was already evident. Output jumped sharply from about 50,000 tons in 1890 to a peak of 133,000 tons in 1916 but then slipped to only 81,000 tons in 1920 (13 percent of U.S. output), heralding the beginning of long-term decline. To be sure, new technologies reduced costs and just barely kept the Michigan mines competitive nationally and on the world scene, but new machines and methods could not overcome the disadvantages of the district's age and geology. Production costs soared as the mine operators went to greater depths only to find poorer deposits.[1]

Forces of Change

In 1865 the abrupt end of wartime demand for copper brought a sharp drop in prices and profits for most of the Michigan mines. Copper prices fell from the peak of 46 cents per pound in 1864 to 36 cents the following year and then to 32 cents in 1866. The decline continued through the end of the decade, reaching a low of 21 cents per pound in 1870. Most Michigan producers cut production and operating costs but were not able to reduce costs quickly enough to keep up with price declines, so profits fell as well. Dividend payments by all Michigan copper mines, which stood at $1,150,000 in 1864, fell to $510,000 in 1865 and reached a low of $100,000 the following year.[2]

The postwar crisis continued in large part because the Michigan copper district increased output substantially. Production fell from 7,179 tons in 1865 to 6,875 tons in 1866, but then nearly doubled between 1866 and 1869, when output reached 13,313 tons. The new Calumet and Hecla mines were responsible for all of the increase. In 1867, the first year of full operation, Calumet and Hecla produced 675 tons of metallic copper, but by 1869 output had reached 6,150 tons, or nearly half the total for the Michigan district. From the late 1860s, the Calumet and Hecla Mining Company was the dominant force in the Michigan copper district for nearly a century. The company's rapid climb to the top is an important chapter in the history of the district.[3]

Edwin J. Hulbert came to the copper district in 1852 to work as a land surveyor. While surveying a road in 1858, he discovered mineral deposits halfway between the Cliff mine and the Portage Lake district to the south, but Hulbert did not uncover the great Calumet Conglomerate Lode until August 1864. Later that year, Hulbert assigned half of the most promising lands to the newly formed Calumet Mining Company and the rest to the Hecla Mining Company. He served as superintendent of the Calumet Mining Company, failed to develop the mine, and speculated disastrously in other mining ventures. By October 1866 he had sold his shares in the Calumet Mining Company to Quincy Adams Shaw, one of the Boston shareholders.[4]

Quincy A. Shaw (1825–1908) had already invested in several Lake Superior mining properties when he first bought shares in the Calumet Mining Company. In March 1867, Shaw convinced his brother-in-law, Alexander Agassiz (1835–1910), to become resident superintendent for the Calumet and Hecla properties. In less than two years after taking charge in March 1867, Agassiz made the Calumet and Hecla mines profitable. He returned to Harvard University in October 1868 and devoted the rest of his life to science, leaving day-to-day operations in the hands of others and visiting the mines only twice a year. The Hecla mine paid its first dividend ($100,000) in late 1869, and the Calumet mine paid the same dividend eight months later. The two companies merged in March 1871 to become the Calumet and Hecla Mining Company. Shaw served briefly as the president of the new firm, but Agassiz then held the post until his death.[5]

The Calumet and Hecla Mining Company (C & H) dominated the Michigan copper industry through the end of the nineteenth century. The firm produced an enormous quantity of copper at costs well below those of the other producers in the district and sufficiently well below market prices to yield healthy profits. The foundation of C & H's success was the richness of the Calumet Conglomerate, which over the period 1871–90, yielded 4.5 percent copper by weight. The Quincy Mine, with some of the richest deposits on Lake Superior, averaged only 2.3 percent copper. The Atlantic Mining Company, south of Portage Lake, earned profits during the 1860s and 1870s with rock containing only 1.1 percent copper.[6]

Calumet and Hecla's costs of production reflected their workers' productivity. In 1871, for example, C & H produced 7,754 tons of metallic copper with a total workforce of 1,225 men, or 6.33 tons per employee. The Quincy Mining Company, one of the more efficient operators, produced 1,205 tons in 1871 with 440 workers, or 2.74 tons per employee. That year, when copper sold for 24.2 cents per pound, costs of production at C & H were 10.5 cents per pound, while the second and third largest producers in the district, the Quincy and Central mines, had average costs of 15.6 cents and 17.0 cents respectively.[7]

Copper production from the Michigan mines stood at 12,311 tons in 1870, doubled by 1880, and then doubled again by 1890, when production stood at 50,705 tons. Technological changes in mining, transportation, and milling, with the mechanization of work previously done by hand, allowed the Michigan copper producers to expand output, lower costs, and remain competitive with the new western mines that opened in the 1880s. Although many of the new technologies seemed minor, their total impact was substantial.

New explosives, for example, replaced black powder for underground use but only after failed efforts to use nitroglycerine oil at the Isle Royale and Huron mines in 1869 and 1870, and a tragic accident at the Phoenix mine in 1874. Dynamite (nitroglycerine oil in an absorbent) became the explosive of choice in the late 1870s for most of the Michigan producers.[8]

The methods used to break rock on the surface before shipping to the stamp mills underwent radical change after the Civil War. Using Eli Whitney Blake's mechanical jaw crusher (1858), most of the Michigan mines built centralized rock houses on the surface between 1864 and 1875. Quincy reduced the cost of breaking rock from 48 cents per ton in 1872 to 30 cents in 1875 as a result of using the new technology. Rock handling methods evolved further in the late 1880s and 1890s, when the major producers incorporated crushing equipment into tall combination shaft-rockhouses, thus eliminating much of the rock handling.[9]

Michigan mines also adopted the pneumatic rock drill in the early 1880s. Beginning in 1868, copper producers experimented with Charles Burleigh's air-powered drill, first used in 1866 in excavating rock for the Hoosac Tunnel in western Massachusetts. Ten of the thirty active Michigan mines used the Burleigh between 1868 and 1873, but it proved to be much too large, unwieldy, unreliable, and uneconomical for use in the copper mines. Several manufacturers tried to sell power drills for mining in the 1870s, but the machines developed by the Rand Drill Company of New York at the end of the decade became the choice of the Michigan mine operators. During the brief span from 1879 to 1883, fifteen of the twenty operating mines adopted the Rand drill, and by 1883 the five largest had 138 drills in use.[10]

The impact of rock drills on output, productivity, and cost was impressive. The Quincy Mining Company's experience shows the results in detail. Between 1877 and 1882, the number of contract miners fell by nearly 40 percent (from 249 to 152), while the volume of ground excavated jumped 22 percent, so productivity doubled. With miners' wages increasing 11 percent in this period, wage costs per ton of rock excavated fell by 44 percent. More important, total operating costs per pound of copper declined by about 20 percent. Other mines reported similar results from the Rand drill. The Osceola, Atlantic, and Pewabic mines reported cost savings for stoping of about 40 percent, while the Franklin claimed savings of 25 percent over hand drilling. C & H increased copper output per miner by 50 percent.[11]

The introduction of the pneumatic rock drill changed the character of underground work in a variety of ways. To help keep the drill operators and their machines fully employed, the mine operators began to use younger boys underground to run a variety of errands. Usually identified simply as "drill boys," the youths kept the miners supplied with sharp drill steel, water, tamping clay, and other supplies. The precise number employed is not clear because the accounts did not always identify them. In the 1890s, for example, the Quincy mine accounts show only 10 to 14 drill boys working when the underground workforce stood at between 400 and 700 men.[12]

Total employment figures for the copper district suggest that the use of younger workers as drill boys had only a minor effect on the use of child labor. A detailed survey of the copper district labor force in 1888 included 2,497 workers, roughly half the total employed in the district. Only 77 boys under 16 years of age (3 percent of the sample) worked in the mining industry, and 62 of them were either 14 or 15 years old. The absence of child labor was even more noticeable in the early twentieth century. The U.S. Census report on mines in 1902 showed a copper district labor force of fewer than 14,000 men aged 16 years and older, with only about 35 boys under age 16 employed.[13]

Additional mines also came into production in Michigan in the mid-1870s. By 1880, four mines that were not in operation in 1865—Calumet and Hecla, Osceola, Allouez, and Atlantic—accounted for 78 percent of Michigan's copper production. The industry became more concentrated over time, with a small number of large producers dominating. In 1865 the ten largest producers accounted for 73 percent of production, but by 1891, the ten largest mines were producing 97 percent of Michigan's copper output.[14]

The Lake producers also achieved greater forward integration into smelting and manufacturing in the late nineteenth century. Following a twenty-year period of increasing concentration of smelting in the hands of outside interests, the major producers built their own smelters. The Waterbury and Detroit Copper Company built a smelter in Detroit in 1850 to treat Lake copper, and in 1860 the Portage Lake Smelting Works opened in Hancock in the heart of the mining district. The two firms merged to form the Detroit and Lake Superior Copper Company in 1867, when most of the remaining smelters that had treated Lake copper had already shut down. Curtis G. Hussey and Company's smelter in Pittsburgh, which operated well into the twentieth century, was the exception. Two small smelters built in the Lake Superior district, at Ontonagon (1863–67) and Lac LaBelle (1866), also failed.[15]

The Detroit and Lake Superior Copper Company, with works at Hancock and Detroit, smelted most of the Lake copper between 1867 and 1887. The Hancock works had eight reverberatory furnaces and three cupolas by the mid-1870s, when the Detroit works had four reverberatory furnaces and two cupolas. In 1875, when output of Lake copper was about 18,000 tons, the Hancock works produced 14,000 tons of ingot copper, and the Detroit works most of the rest. The 1880 Census identified only six eastern copper smelters, including the two works of the Detroit and Lake Superior Company, at Detroit and Hancock; Hussey's Pittsburgh plant; the Baltimore Copper Company's works in Baltimore; and smelters in Vermont and Tennessee to process the ores produced in those states.[16]

Calumet and Hecla was the first Lake producer to control the smelting of its own copper. After C & H threatened to build its own smelter, the Detroit and Lake Superior Copper Company agreed in 1885 to jointly build and operate a smelter with C & H on Torch Lake.

The smelter site, originally named Grover, became South Lake Linden in 1889 and has been named Hubbell from 1903 to the present. The Detroit and Lake Superior Copper Company, however, had to agree to let C & H buy out the smelting company's shares in 1892. After the smelter went into operation in 1887, the Detroit and Lake Superior Company closed its Detroit works. In 1892, with five years' operating experience under its belt, C & H bought the remaining shares and opened a second smelter at Buffalo, New York.[17]

The Quest for Stability

The drop in copper prices following the Civil War, which reflected declining demand and increased foreign competition, persuaded copper mine operators in Michigan to seek greater tariff protection. The tariff act of 30 June 1864 set duties on ingot copper at 2.5 cents per pound; 35 percent ad valorem on manufactured goods; and 5 percent ad valorem on copper ore. Representative John F. Driggs of Saginaw, Michigan, opened the new debate on tariff protection on 22 June 1865 when he proposed an increase in duties to 6 cents per pound on ingot copper and 3 cents per pound on the copper content of foreign ores, the equivalent of a rate of about 40 percent ad valorem. A United States Revenue Commission established in 1865 to examine changes in postwar tariffs and excise taxes began hearings in New York City in November 1865 on the state of the American copper industry. The arguments for and against increased protection offered in the commission hearings then reappeared repeatedly throughout the resulting debate.[18]

The real distress faced by the Michigan copper mines, along with their political clout, eventually won the day. About twenty mines produced copper in Michigan during the immediate postwar period, but only three of these paid a total of six dividends in 1866, 1867, and 1868. The next two years brought only slight improvement, with three operators (Calumet and Hecla, Quincy, and Central) paying dividends in both years. The advocates of protection naturally exaggerated the copper industry's depressed state. In late December 1868, Michigan senator Jacob M. Howard argued that only eleven mines were working in Michigan, whereas more than eighty had been operating at the end of the Civil War. He also claimed that the Lake Superior district had more than fifteen hundred boarded-up mine houses, abandoned by families left unemployed by the mining companies. Howard's accounts of conditions in the Michigan mining district were pure fiction.[19]

The tariff bill that reached Congress in early 1869 provided for a 5-cent duty on ingot copper and a 3-cent duty on the copper content of ore. Given the depression in the copper industry and the fact that other metal industries already enjoyed substantial tariff protection, both houses passed the bill on 8 February 1869. President Andrew Johnson vetoed the measure on 23 February, but Congress overrode his veto the next day. The 1869 tariff increased duties on "old copper" (scrap copper) from a half cent per pound to 4 cents, doubled the duties on copper imported as pigs or bars (from 2.5 cents per pound to 5 cents), and increased the rates on manufactured copper from 35 percent to 45 percent. The immediate impact was small, because imports of old copper and pigs had already peaked in 1866 at 1,520 tons and then had fallen sharply by 1868 to 174 tons, less than 2 percent of American output.

The increased duties on ore devastated the East Coast smelting industry. To be sure, the smelters faced a cost squeeze even before the new tariff because the copper content of imported ores was declining. The Bergen Point smelter in New Jersey, also known as the Soho or Belleville works, had already shut down in 1866. The Point Shirley works in Boston smelted domestic ores until 1872, when it shut down permanently. The Baltimore and Cuba Mining and Smelting Company works at Locust Point in Baltimore closed in 1868 to make way for railroad yards, and the firm, reorganized as the Baltimore Copper Company, moved all production to its Canton works in Baltimore. The Canton works temporarily closed in 1869 because of the tariff but reopened in the early 1870s as new ores from Arizona and Montana came onto the market. The smelter continued operating into the 1890s as the Baltimore Copper Smelting and Rolling Company.[20]

The impact on the eastern smelters was particularly devastating because the tariff dried up foreign ore supplies at the same time that most domestic ore supplies temporarily disappeared. The Bristol (Connecticut) Mining Company had ceased production before the

Civil War, while copper mines in Vermont and Tennessee had their own smelters. A half dozen new copper mines that opened in northern California during the Civil War only briefly supplied the eastern smelters with ore. The first significant California shipments to eastern smelters, a total of 3,660 tons in 1862, had nearly tripled by 1864, when exports from San Francisco to the East Coast reached 9,970 tons. A half-dozen mines at Copperopolis in Calaveras County northeast of San Francisco accounted for most of the 68,631 tons of ore and 850 tons of regulus shipped from San Francisco to East Coast and Welsh smelters between 1862 and 1867. About three-quarters of the ore shipped from San Francisco, or about 50,000 tons, went to East Coast smelters. There is no reliable evidence about the copper content of these ores, but Browne's report on California ore shipments suggests that the metallic content had to be at least 15 percent for mine operators to ship the ores profitably. If the average copper content was about 20 percent, shipments of 50,000 tons of ore would yield about 10,000 tons of metallic copper. The California mines quickly shut down in late 1866 and early 1867, victims of the sharp decline of copper prices following the end of the war. By the time of the Census of 1870, none of the California mines were at work.[21] The dwindling supply of ore for the eastern smelters is seen in table 3.1.

The large increase in imports in 1869 was nothing but a last-minute effort of suppliers and consumers of foreign ores to beat the increase in duties. Imports during the last four months of the year ending 30 June 1869, after the new duties were in place, amounted to only 88 tons. The effective rate of duties on imported ore jumped from 5 percent to 43 percent as a result of the new tariff legislation.[22]

The passage of the tariff of 1869 did not bring an instant return to prosperity for the Michigan mines, as prices continued to plummet. Market prices of Lake copper in New York, which stood at 26 cents per pound in February and March 1869, dropped disastrously to only 19 cents by May 1870 and averaged under 21 cents for the entire year. The leading newspaper of the Michigan district, the *Portage Lake Mining Gazette,* noted in June 1870 that five area mines were about to shut down and that only five producers could even hope to earn profits at current copper prices. A federal government decision to dump 5,000 tons of war-surplus copper on the market (equal to more

TABLE 3.1
Copper Ores Shipped to East Coast Smelters, 1865–1870

	From San Francisco[a]	From Foreign Sources[a]	Total[a]
1865	13,196	9,282	22,480
1866	14,499	3,072	17,571
1867	2,633	15,300	17,993
1868	0	6,143	6,143
1869	0	11,088	11,088
1870	0	381	381

Sources: John Ross Browne, *Report on the Mineral Resources of the States and Territories West of the Rocky Mountains* (Washington, D.C.: USGPO, 1868), 218; and *Reports on the Commerce and Navigation of the United States*, House Executive Docs., 39th, 40th, and 41st Congresses. The import statistics are for the year ending 30 June.

[a]Tons of 2,000 pounds.

than a third of the industry's annual output) only exacerbated the producers' problems. The mine operators sought solutions to the low price levels by establishing pooling arrangements that featured efforts to dump excess supplies overseas.[23]

The principal copper producers, including the eastern smelters, met in New York and Boston in early March 1870 to devise strategies to end the decline of copper prices. The first meeting, on 1 March, included representatives of eight Michigan mines. They agreed to export at least 5 million pounds to help clear the domestic market and appointed a committee to work out the arrangements. At the next meeting, on 10 March, the producers confirmed their agreement and allocated sales among the participants. At their final meeting in New York on 12 March, the companies made only minor changes in the agreement.[24]

The Lake copper producers, with C & H in the lead, tried to stabilize the market through various domestic pooling arrangements as well. In 1872, C & H reached an agreement with N. S. Simpkins Jr. of New York by which C & H would consign its entire output for two years to Simpkins for sale in the U.S. market. The New York brokerage firm of Holmes & Lissberger tried to corner the copper market in early 1874, when it offered to buy the entire output of the Michigan mines at an above-market price and dispose of any surplus overseas. The firm went into bankruptcy later in the year, however, and the corner collapsed.[25]

These efforts to sell surplus output overseas produced some price relief and helped stabilize the market. The mine operators exported 7.2 million pounds (net of imports) in 1870 and 8.3 million pounds net in 1871. The average selling price of copper in 1871 was 22.6 cents per pound, 2 cents above the 1870 level. The Lake producers continued pooling for export sales from 1875 on, but we know little about the internal workings of the pool. They appear to have worked together to stabilize copper sales on the domestic market as well. Calumet and Hecla and the other major producers routinely developed long-term contracts with major industrial customers and sold to large metal brokers abroad, so the mining companies sold little copper in the open market.[26]

The combination of the 1869 tariff, selective dumping overseas, and restraints on domestic production allowed American producers to enjoy a price premium of at least 2 cents a pound over London prices in the 1870s. American producers exported nearly 94 million pounds of copper between 1870 and 1880, one-fifth of total output. U.S. copper output doubled between 1870 and 1880, but prices stayed above 20 cents a pound except for the three-year period starting in 1877. The pooling/dumping arrangement was manageable and effective because the Lake mines accounted for three-quarters of U.S. output and Calumet and Hecla produced two-thirds of the Lake total.[27]

In December 1882 the major Lake producers contracted with N. S. Simpkins Jr. to sell all their unsold copper, a total of 15.7 million pounds, while agreeing to ship no more copper out of the mining district until the opening of navigation in 1883. By the end of March 1883, the same producers had crafted another agreement, which replaced the December 1882 accord. The new pact, to run to the end of 1883, empowered Simpkins to sell 50 million pounds of copper, roughly the yearly production of the signatories. C & H signed separate pooling agreements with Simpkins, with the last pact extending until 31 March 1886. The C & H pooling arrangements for the years 1883–86 covered all sales, both foreign and domestic.[28]

In the early 1880s, C & H had increasing trouble controlling copper prices because of increased production from mines in Arizona and Montana. Production from outside of Michigan increased from 10 million pounds in 1880 to 55 million in 1883 and to 76 million the following year. The Lake companies did not include the new producers in their agreements, which guaranteed failure. Prices fell from 20 cents in 1880 to under 16 cents in 1883 and then to 14 cents in 1884. The Quincy Mining Company was the first major producer to abandon the cartel. It concluded that its involvement in the pool was no longer in Quincy's best interests and began foreign sales on its own in November 1884. The other pool members sought an injunction in the Michigan courts, but Quincy successfully argued that the pool operated not in the public interest.[29]

The efforts of Calumet and Hecla and the other Lake Superior producers to bring order to a chaotic market failed because they could not control production. American copper output increased from 60 million pounds in 1880 to 166 million in 1885, while prices fell from 20 cents a pound to 11 cents. Calumet and Hecla and the other Lake producers deliberately slashed prices in the early 1880s to drive the Butte, Montana, mines out of business, but they were ineffective because by the mid-1880s they were producing only half the national output. Calumet and Hecla increased production in 1886 and 1887 and drove down prices for large contracts to 10 cents a pound. However, underground fires at the Calumet and Hecla mines in 1887 reduced output below the level of 1886 and ended the firm's efforts to flood the market. Low prices forced several of the Montana mines to close in 1886 and 1887, but the Anaconda remained open, and by the fall of 1887 the Lake producers had lost control of the American market.[30]

In a fortunate turn of events for the U.S. mine operators, Hyacinthe Secrétan, director of the Société Industrielle et Commerciale des Métaux, the largest European copper manufacturer, began an effort in October 1887 to control the supply of copper worldwide. The Secrétan Syndicate agreed to buy the entire output of the major producers, with restrictions on output, and would then control global sales. Betting on future increases in demand and prices, the syndicate offered producers between 13 and 15 cents a pound when world prices stood at 11 cents. By March 1888, Secrétan had signed three-year agreements with the major American producers except for the Anaconda, which had two six-month agreements covering 1888. Similar agreements with the giant Rio Tinto mine in Spain and the principal mines in Chile and South Africa gave the Secrétan Syndicate control over roughly 85 percent of world copper production.[31]

Secrétan succeeded in artificially raising copper prices, but world supply increased and demand had leveled off, leaving the syndicate with large stocks of surplus copper. The price of Lake copper at New York, an anemic 11 cents in 1887, jumped to an average of over 16.5 cents for 1888 and the first four months of 1889. The syndicate had bought 232,000 tons of copper during 1888, mostly on credit, but sold only 87,000 tons. When copper prices slumped in March 1889, the syndicate went bankrupt.[32]

The potentially devastating impact on copper markets never occurred, because the Syndicate sold its copper stockpiles gradually and global demand increased substantially after the collapse. The Rothschilds, the syndicate's greatest creditor, agreed to release their stocks over a four-year period, after the two largest American producers—Anaconda and Calumet and Hecla—threatened to flood world markets with cheap metal. Lake copper prices fell to 11 cents in September and October 1889 but recovered by the end of the year and in 1890 averaged nearly 16 cents a pound.[33]

The domestic copper industry enjoyed effective tariff protection for the period from 1869 to 1890. In 1890 the McKinley Tariff brought an end to protection by lowering the duty on copper ore to 0.5 cents per pound and that on ingot copper to 1.25 cents. The tariff act of 1894, the Wilson-Gorman Tariff, eliminated duties on ore, regulus (partially smelted ore), scrap copper, and ingot copper, while reducing duties on copper manufactures to 35 percent ad valorem. The items placed on the free list in 1894 remained duty-free until the depression of the 1930s. The mining companies made no serious effort to keep the protective tariffs in place.[34]

Management and Labor to 1890

Labor-management relations were uneventful, if not always friendly, during the quarter century following the Civil War. Conflicts over wages, hours, and working conditions occurred, but the small size of most mines allowed easy, informal communication between managers and workers, making nonviolent conflict resolution feasible. Calumet and Hecla was the single exception to the rule, with 1,200 employees by 1870 and 3,500 in 1890, easily one of the largest mines in the country. The Quincy, which had a peak employment of 654 workers in 1865, had a workforce of between 400 and 500 in the 1870s and 1880s. The Atlantic mine was in the same size range as Quincy during the 1880s, when the Central and Allouez mines had between 200 and 300 employees. Finally, two early giants in the Michigan district, the Cliff and the Minesota, each had fewer than 200 employees by the mid-1870s and fell below 100 in the decade that followed.[35]

The small size of the mines resulted in a small and unarticulated management structure that almost guaranteed a face-to-face, informal managerial style. In June 1865, when the Quincy Mining Company had a total of 654 employees in Michigan, the entire management team, including bosses and foremen, amounted to twenty-five men. Twenty years later, with a reduced workforce of exactly 400 men, the number of managers had fallen to eighteen. Direct, informal communications between the miners and the mine superintendent, including airing of grievances, were frequent as long as the mines remained small. In many instances, small delegations of workers, usually miners or other skilled men, met with the management and settled conflicts without resorting to strikes.[36]

Some strikes did occur during this period, however, and mine managers were concerned over the possibility. Significant strikes took place at the nearby iron mines in Marquette County in July 1865 and June 1866. Trammers at the Pewabic mine walked out briefly in April 1866 in a failed attempt to increase their wages. To prevent or defeat future strikes, seven agents representing ten mines established the Houghton County Mine Agents Union in March 1869. They agreed to keep each other informed of active strikes, but more important, they also promised not to hire any new employees while a strike continued at any of their properties. They would not take advantage of a struck mine by increasing output and would not relieve the economic distress faced by striking mine workers. The agents met five times between March and June 1869 but then disbanded.[37]

One labor dispute exploded at the Huron mine in mid-April 1870 after the owners invited the Mabbs Brothers Company to experiment with nitroglycerine oil as a substitute for black powder. The Huron miners refused to work or to allow the experiment to proceed, believing that using nitroglycerine would endanger their lives. The "Glycerine Strike" continued for a week until the men ignited a keg of black powder next to the entire stock of nitroglycerine oil in a shack on the surface, blowing everything to smithereens. The Huron Mining Company then canceled its contract with the Mabbs Brothers. Less than a year later the *Portage Lake Mining Gazette* urged the mine operators to adopt the safer explosive, "Giant Powder," a type of dynamite, as a significant economy measure, but the legacy of the Huron strike slowed its use until the second half of the decade.[38]

The only districtwide strike before 1890 began at Calumet and Hecla on 1 May 1872. A sharp rise in the cost of living in the copper district in early 1872, combined with shortages of miners, brought demands from the men at Calumet and Hecla for substantial wage increases. R. J. Wood, the superintendent at C & H, distributed a circular dated 27 April in which he offered his workers an immediate 10 percent pay increase and an intriguing profit-sharing plan. On 1 January 1873, employees who had worked for the company for twelve months would split a sum consisting of 1 percent of the revenues from a monthly product of 600 tons of ingots and 2 percent on output above 600 tons, with a guaranteed minimum of $50,000 to be divided in proportion to the employees' wages. The C & H underground workers refused to work on 1 May and quickly forced timbermen and surface workers to leave the mine. Similar strikes began at the other Houghton County mines—including the Schoolcraft, Quincy, Pewabic, and Franklin—with the men demanding wage increases and a reduction of the workday from ten hours to eight. On May 8 the C & H superintendent asked Michigan's governor for troops to protect lives and company property.[39]

The official request for military assistance came via a telegram from Bartholomew Shea, the Houghton County sheriff, to Governor Henry P. Baldwin. Shea requested two companies of infantry and suggested they come from Chicago to hasten the restoration of civil order. Baldwin requested federal troops from Chicago, but General Phillip H. Sheridan, the renowned Civil War general and now commander of the Military Division of the Missouri, informed Baldwin that Houghton County, Michigan, was outside of his jurisdiction and suggested that Baldwin contact General Cook in Detroit. Besides, Sheridan had no troops available because they were all off serving in Utah.[40]

Conditions in Calumet quickly deteriorated further. On May 13, Shea arrested two of the strike ringleaders and was escorting the men to jail when a mob of about six hundred strikers overpowered the sheriff and freed the prisoners. Sheriff Shea repeated his request for military assistance on May 13, this time asking for a minimum of three companies. General Cook sent two companies on the steamer *Atlantic* on May 12 and an additional company by rail via Chicago and Green Bay on May 14, but the first troops did not reach Portage Lake until late in the afternoon of May 16.[41]

The arrival of federal troops quickly brought the strike at Calumet and Hecla to an end. Miners at the other mines had already settled in mid-May for increases of about 10 percent, and the C & H workers followed soon after, settling on May 22 for a 10 percent increase but with a ten-hour day. As the troops were about to leave the copper district, several iron mine operators in Marquette County asked the governor to leave one company of soldiers in Marquette or at Sault Ste. Marie for sixty to ninety days because "they will no doubt exert a good influence upon our labor, & may prevent any trouble with our miners."[42] Their request was denied.

The mine operators in the copper district faced continued labor unrest through early 1874. Calumet and Hecla's trammers struck on 1 January, but the miners did not join them. The dispute lasted less than a week, when the men returned to work.[43] In the aftermath of this strike, President Agassiz summed up the owners' attitude toward labor in a letter to the C & H superintendent:

> We cannot be dictated to by anyone. . . . As I have written to you before we have always treated our men fairly and honestly, they have received higher wages than [from] any other corporation. I have attempted formerly to try and get their good will by offering them a share of the profits. They spit in my face as it were and all we can do is to sit quietly and await results. Wages will be raised whenever we see fit and at no other time.[44]

The Michigan mine operators had no recorded labor disturbances between January 1874 and June 1890. Calumet and Hecla, however, suspected that disgruntled workers had set several costly underground fires in the 1880s, and the company reacted quickly to Knights of Labor organizing drives between 1887 and 1890. The company hired detectives to provide intelligence, fired men suspected of K. of L. sympathies, and briefly established a company union, but the Knights were never a real force in the copper district.[45]

Michigan Copper in the Larger Context

The production of copper in the United States and throughout the rest of the world underwent dramatic growth and geographic shifts during the quarter century following the American Civil War. The American copper industry became increasingly important in world copper markets, while Michigan's dominant position within the industry declined sharply starting in the early 1880s. The growth of world copper output and America's place in that growth can be seen in table 3.2.

The world copper industry outside of the United States underwent drastic changes in the late nineteenth century. The British (Cornish) copper industry practically disappeared, with output of only 1,000 tons in 1890, making Britain a minor producer on the world scene. Chile, which accounted for roughly half of world output in 1860, achieved peak production of about 50,000 tons in the late 1870s before experiencing a gradual decline in production to about 25,000 tons by the early 1890s. By 1900, Chile's share of world production was only 5 percent. The mines of Spain and Portugal became the most important new source of copper outside of the United States in the late nineteenth century. Their metallic copper output reached 60,000 tons by the early 1890s, with a single producer, the Rio Tinto, accounting for more than half the total.[46]

The extraordinary growth of U.S. output during the 1880s reflected the growth of copper consumption in the United States. Domestic consumption, about 12,000 tons in 1870, more than doubled to 27,000 tons in 1880 but then nearly quadrupled to about 95,000 tons by 1890. Per capita consumption of copper in the United States, about

TABLE 3.2
World and United States Copper Production, 1860–1890

	World[a]	U.S.[a]	U.S. Share
1860	95,000	8,000	8.4%
1870	105,000	14,100	13.4
1880	172,000	30,200	17.6
1890	302,000	130,000	43.0

Sources: Michael G. Mulhall, *The Dictionary of Statistics* (London: Routledge, 1892), 156; *Mineral Industry, 1892,* 118; and William B. Gates Jr., *Michigan Copper and Boston Dollars: An Economic History of the Michigan Copper Mining Industry* (Cambridge, Mass.: Harvard University Press, 1951), 197–98.
[a]Tons of 2,000 pounds.

one pound in 1880, exceeded three pounds in 1890 and reached four pounds in 1892. Much of this extraordinary increase in demand was a result of the growth in demand for electricity and the resulting demand for copper wire.[47]

Michigan's mines dominated the U.S. copper industry from the late 1840s until the early 1880s, when the enormous deposits of Montana and Arizona came into the market. During 25 of the 34 years between 1847 and 1880, for example, Michigan produced more than three-quarters of the American output. Lake Superior's share of U.S. production increased from 77 percent in 1845–1854 to 84 percent in 1871–1880. Michigan's share of the total was lower in the earlier period because during the years 1862 to 1866 it produced only two-thirds of the total.[48]

The Lake producers faced few serious competitors between the late 1840s and 1880. One exception, discussed earlier, was the substantial but short-lived copper mining district that emerged in California during the Civil War, a hothouse industry that quickly withered and died when copper prices collapsed.

The Ducktown, Tennessee, mines produced substantial amounts of copper briefly in the 1850s and contributed to the Confederate war effort through 1863, when federal forces disrupted operations there. The Ducktown district, also known as the Appalachian Copper Basin, rebounded after the war, with a peak output of about 1,200 tons in 1878. Ducktown's best performance, however, amounted to less than 5 percent of the national output. The dominant firm, the Union Consolidated Mining Company, went into receivership in 1879, and the Ducktown mines remained closed until 1893. The district revived in the early twentieth century, when it produced copper in modest quantities.[49]

Vermont was the only other copper region of any importance before 1880. The Copperfield or Ely mine, also known as the Vershire, was the largest and most consistent producer from 1854 through 1884, when it went into receivership and closed permanently. Three others—the Elizabeth, Union, and Eureka mines—enjoyed only spotty success. Vermont was the second largest producer, with about 7 percent of national output in 1870 and roughly 5 percent in 1880. A decade later, however, production from all the eastern and southern states was about 4.2 million pounds, only 1 percent of the U.S. total.[50]

The Lake copper producers dominated U.S. production until the 1880s, when their position quickly and permanently changed with the appearance of western copper. Michigan's share of national output dropped from 82 percent in 1880 to 39 percent in 1890 and to only 24 percent in 1900. Prospectors discovered copper in Arizona in 1877 and 1878, but serious mining did not begin there until 1880. Arizona Territory produced about 3 million pounds of ingot copper in 1880, when national output was 60 million pounds. Production then skyrocketed, and from 1882 through 1884 Arizona produced one-fifth of American output. Its share of national production fell below 10 percent in 1886–87 but then rebounded at the end of the decade.[51]

The most significant development of the 1880s was the exploitation of the rich copper ore deposits at Butte, Montana. The first copper smelter at Butte dated from 1879, but output in 1880 was a mere 600 tons. The discovery of the enormous deposits at the Anaconda mine in 1882 marked the beginning of the end of Michigan's dominance. As early as 1882, Montana produced 10 percent of American copper, and by 1885 it was accounting for 41 percent and nearly equalled Michigan's total. Montana's copper output surpassed Michigan's in 1887, and Butte became the dominant mining camp in the country, a position it kept well into the twentieth century. The emergence of the western copper mines and their place in the U.S. copper industry are considered in the next chapter.

Management and Labor, 1890 to 1920

The labor peace that had prevailed since 1872 ended abruptly with a spate of strikes in the spring and summer of 1890. Chronic labor troubles then plagued the district through the great strike of 1913–14. Each strike was the result of particular grievances—including wages, hours, and working conditions—but more systemic forces were at work as well.

The mining companies were much larger than before, and their greater size contributed to labor unrest. The Quincy Mining Company, for example, employed only 400 workers in 1885 but by 1905 had more than 1,700, and in 1909 reached a peak employment level of 2,045 in its Michigan operations. Calumet and Hecla, which already employed more than 2,000 in 1885, had a workforce of more than 5,000 at the turn of the century and in 1917 had a total of 6,017 employees, the company's all-time peak. At the onset of the strike of 1913–14, the Copper Range Consolidated Company had 2,716 employees, the Osceola Consolidated Mining Company employed 1,143, and four other mines (Mohawk, Ahmeek, Isle Royale, and Tamarack) had workforces ranging between 610 and 851 men. The eight largest mines employed 87 percent of the labor force. The typical Michigan copper worker in the early twentieth century was part of a labor force of more than 600, and frequently well over 1,000.[52]

The working relationship between mine managers and their employees changed considerably during this period, in part because the mining companies, and the management structure required to run them, had grown in size and complexity, widening the physical, social, and economic gap between management and labor. The personal, informal lines of communications that had existed when mining companies were small broke down when they became larger. The greater size and impersonality of the mining operations also diminished management's ability to gather information on a variety of subjects, including workers' sentiments.[53]

Most of the strikes that took place between 1890 and 1913 affected only one or two mines in the district. Simultaneous strikes at enough properties to constitute a districtwide disturbance occurred only in 1890, 1904, and 1913–14. A wave of strikes involving both underground and surface workers swept through the district in late June and early July of 1890 and involved the Atlantic, Calumet and Hecla, Franklin, Kearsarge, Osceola, Quincy, and Tamarack mines. Quincy's workers settled for a general pay increase of 10 percent, and other workers received similar settlements. When the first ripples of the strike wave arrived at the Tamarack on June 20, Allen F. Rees, the prosecuting attorney for Houghton County, warned Michigan's governor Cyrus G. Luce that troops might be needed to preserve civil order. Tamarack, however, quietly gave its employees a 10 percent increase and the strike ended peacefully.[54]

The scattered strikes of the 1890s and early 1900s had various root causes but were often defensive in nature—that is, they were called to prevent pay cuts or increased workloads. The miners themselves rarely initiated strikes, although they sometimes supported the actions of other workers. Trammers (unskilled workers, most of whom were recent immigrants from Austria-Hungary, Finland, Italy, or Poland) spearheaded most of the work stoppages after 1890. They struck the Osceola mine in June 1892, Calumet and Hecla in May 1893, and Quincy in April 1896, and they threatened to strike at a half-dozen other locations. In January 1904, Italian trammers supported by their Finnish co-workers initiated a serious strike at Quincy, which closed the mine for two weeks.[55]

The Michigan copper district experienced considerable labor unrest between 1904 and 1906, with escalating stakes and greater violence evident, presaging the events of 1913–14. The changing character of labor relations in the industry can be seen in two labor disputes in the summer of 1906. The first began on 23 July at the Quincy mine, when the underground workforce walked out, demanding a 10 percent pay increase. The strike continued for three weeks and ended when most of the men accepted the company's first and only offer of a 5 percent raise. This strike was significantly different from previous disputes in that the men formed a multi-ethnic strike committee and presented the company with a set of seven written demands. Besides the pay raises, Quincy's workers asked for a one-year agreement, no recriminations against strikers, and improved working conditions, including better enforcement of safety regulations.[56]

The second dispute involved a group of Finnish trammers who struck the Michigan mine at Rockland in Ontonagon County. The Finnish White Rose Temperance Society of Rockland had invited a Western Federation of Miners organizer to speak to the mine employees on the Fourth of July, and the organizer had urged the men to strike for higher wages. The Western Federation had attempted to organize Michigan copper workers starting in 1903 but had achieved little success.

The Finnish workers at Rockland struck for higher wages on 30 July 1906. Ontonagon County sheriff William McFarlane and a small force of deputies arrived on the scene the same day. During a confrontation on 31 July, the forces of law and order killed two strikers, wounded nine others, and arrested more than a hundred men. The events at Rockland served as a preview of the greater drama played out seven years later. The Rockland violence also became an international incident, although a minor one. Several of the Finnish laborers at Rockland formally complained (as Russian citizens) to the Russian ambassador to the United States about their treatment and asked him to intervene on their behalf. There is no evidence that he did anything.[57]

The districtwide strike that began on 22 July 1913 under the banner of the Western Federation of Miners threatened the system of labor relations, including company paternalism, used by management since the beginning of mining on Lake Superior. There is no room here to tell the full story of this monumental struggle, including its many dramatic and tragic events, but the major developments of the strike were as follows.[58]

The Western Federation of Miners (WFM), after failed attempts to organize Michigan's copper miners between 1903 and 1906 and a disastrous strike of Mesabi Range iron miners in 1907, returned to the Michigan copper district in late 1908. The WFM had established local unions in Calumet, Hancock, South Range, and Mass City by 1910, when the union had over 800 members in the district. Even in the early months of 1913, the miners' union had enrolled fewer than 1,000 men, when the mines employed about 14,000 men districtwide.[59]

The WFM employed immigrant organizers to appeal to the various ethnic groups, and by 1 June 1913 nearly 3,300 men had joined the union. Membership ballooned to more than 7,000 by early July, when the locals voted in overwhelming numbers to authorize a strike if the mining companies refused to negotiate with them. The WFM's national officials in Denver wanted the Michigan locals to build their membership further before initiating a strike, but the Denver officials lost control of events and the strike began on 22 July.[60]

The WFM made five demands of the mine operators: an eight-hour day; a minimum wage of three dollars a day for underground workers; abolition of the one-man drill; a formal grievance procedure to settle workers' complaints; and recognition of the Western Federation of Miners. The first two demands were at least discussed by the parties to the conflict, whereas the mining companies would not even consider the last three. The WFM argued that the mines should reduce the workday from an average of nine hours to the eight-hour standard prevailing in Butte and in Michigan's iron mining districts. Calumet and Hecla's miners earned an average of about $3.30 a day, and its trammers about $2.75, but at the other mines in the district the comparable wages were $2.75 and $2.40 respectively. Miners' wages in the Copper Country were comparable to those paid in the iron mining districts in Michigan and Minnesota, but the Butte, Montana, mines consistently paid 50 to 60 percent more.[61]

The mine operators presented convincing evidence that the cost of living in Michigan, especially rent and the cost of provisions, was lower than in Butte. When company-provided benefits and amenities such as medical services, accident and retirement programs, libraries, and the like are included, Michigan mine employees enjoyed a higher standard of living than their counterparts in Butte. The mining companies also argued that a standard eight-hour shift was impractical. The lengths of shifts varied greatly within the district, from nine to ten and a half hours, because of the varying amounts of time needed to reach the work area underground. Time spent at work averaged between eight and nine hours per day.[62]

Elimination of the one-man drill was a surprising demand in many respects. The mine operators claimed that they had to use the new technology if they wished to remain competitive with the western mines. The new drills were particularly valuable in allowing the exploitation of the narrower veins the mine operators encountered at greater depths. Besides, the companies gave miners a substantial pay raise when they switched from the two-man to the one-man machines. The miners complained that while one man could operate the drill by himself, he needed help to set it up, but mine managers had made no provision for assistance.[63] More important, miners argued that the new machine was dangerous because a man working alone could suffer an injury—from a cave-in, for example—and die because nobody would know of his predicament. The miners called the new machines "man-killers" and "widow-makers." A new state law about to go into effect on 12 August 1913 requiring that miners operating one-man drills work no more 150 feet from the nearest fellow worker did not quell the miners' fears. For a miner stuck in the dark, twisting, noisy labyrinths of underground workings, 150 feet seemed more like a mile. The one-man drill remained a key issue for the strikers, both symbolic and real.[64]

Evaluating the issues of the strike is difficult. A month after the strike began, the mine managers announced that they would introduce an eight-hour day, with a three-dollar minimum daily wage for underground workers, on 1 January 1914 but insisted that they had planned those changes all along. At least one historian has argued that the abusive treatment of underground workers by petty bosses, combined with the lack of mechanisms for resolving employee grievances, were the more fundamental complaints. Michigan's copper mine workers felt powerless in their relationship with the mining companies, especially with the giant Calumet and Hecla. The strike was about dignity and power, not bread and butter or one-man drills.[65]

In the early phases of the strike, the WFM was able to shut down the major mines. After several days of rioting and violence, Governor Woodbridge N. Ferris sent the entire National Guard of Michigan, some 2,600 men, to the Copper Country to restore law and order, but he had withdrawn more than half of them by late August. The mine operators did not attempt to resume even limited production before early September and had little success doing so until they began importing strikebreakers in the middle of the month. Quincy was the first to resort to this practice, and by the end of 1913 the company had brought in about 525 new men. Calumet and Hecla had imported close to 2,000 strikebreakers by the end of the year.[66]

The conflict turned into a costly, often violent endurance contest, which the mine operators were decidedly winning by the end of 1913. Despite dozens of injunctions, lawsuits, attempts by various parties to mediate a settlement, investigations, and inquiries, including a three-month congressional investigation that started in January 1914, the strike remained a head-to-head power struggle that the mining companies won because they accepted huge short-term losses in exchange for victory. The WFM spent about $800,000 in the struggle, but in April 1914 it held a referendum on the issue of continuing the strike. By a 4:1 margin, the 2,400 remaining union members voted to give up, and on 14 April 1914 the WFM declared the strike officially ended.[67]

The Michigan copper district faced labor shortages through most of 1917 due to the loss of manpower to the war effort and the appeal of high-paying factory jobs in cities like Detroit and Chicago. Quincy hired 2,525 new employees in 1917 but actually increased its workforce by only 90 men. The Pinkerton Agency screened new hires to ensure that the mine operators did not inadvertently hire Industrial Workers of the World (IWW) activists or sympathizers. The IWW appeared in the Michigan copper district in July 1917 and launched an abortive strike, but it had little impact.[68]

Expansion, Technological Change, and Rising Costs

Michigan's copper output grew impressively after 1890 but more slowly than production from the other American mining districts. Although the product of the Lake mines nearly trebled between 1890 and 1916, jumping from about 51,000 tons to 135,000, Michigan's share of national output fell from 38 percent to 13 percent over those years. Most of the increased production in the Lake district came from new mines. In 1890 the four largest mines--Calumet and Hecla, Tamarack, Quincy, and Osceola—accounted for 86 percent of production, but they increased output by only one-third between 1890 and 1916. In the latter year they accounted for only 44 percent of Michigan's output.[69]

Nearly half of the district's output in 1916 came from deposits unknown or unused in 1890. Two substantial mines, the Mohawk and the Ahmeek, worked the northern extension of the Kearsarge amygdaloid lode discovered in 1896 in southern Keweenaw County. Together they produced about 19,000 tons in 1916, or 14 percent of the district's total. The Baltic or South Range Lode, located south of Portage Lake and discovered in 1897, was even more significant. By 1916, four new operations—the Baltic, Champion, Superior, and Trimountain mines—produced about 28,900 tons of ingot copper, more than 21 percent of the district total. Two additional mines, the Allouez and the Isle Royale, barely operating in 1890, produced 11,300 tons in 1916, about 8 percent of the total for the Lake Superior district.[70]

The large producers that dominated the Michigan copper industry in the early twentieth century were the product of the financial and operational consolidation of about two dozen properties. Quincy aggressively bought neighboring mines and real estate, including the Pewabic (1891), the Pontiac and Mesnard (1896), the Arcadian (1906), the Franklin (1908), most of the Hancock (1915–1919), and other adjacent mineral lands. The Osceola, controlled by an alliance of Albert Bigelow and Leonard Lewisohn, merged with the Iroquois, Tamarack Junior, and Kearsarge mines in 1897 to form the Osceola Consolidated Copper Company. The third major group of mines included eight properties controlled by John Stanton and the Copper Range Consolidated Company, a holding company that Stanton and William A. Paine established in 1901 and that owned the Baltic, Champion, and Trimountain mines. The Calumet and Hecla Mining Company was the fourth major producer. In 1904, these four producers controlled 96 percent of Michigan's copper output.[71]

Calumet and Hecla also moved to consolidate its operations with those of its neighboring mines. In 1905, Michigan began to permit mining companies to own stock in other mines, and Calumet and

Hecla began to purchase a controlling interest in several adjacent mines, including the Osceola. In early 1907, C & H claimed to own most of the Osceola stock, but Albert Bigelow stopped it from assuming control by obtaining a U.S. Circuit Court injunction preventing the takeover. In early 1909, as the case was slowly moving through the courts, Bigelow sold all his stock in Michigan mines to C & H for more than $8 million. Calumet and Hecla took control of these properties—including the Osceola, Ahmeek, Isle Royale, and Tamarack—in 1909 and 1910 and proposed a complete consolidation of all operations the following year. A series of minority stockholders' lawsuits brought in the Michigan courts, however, forced Calumet and Hecla to abandon its consolidation plan.[72]

The copper companies viewed expansion and consolidation as a way of reducing the escalating mining costs that resulted from increased working depths and falling yields from the rock mined. By the turn of the century, Calumet and Hecla's vertical Red Jacket Shaft reached 4,920 feet, while the nearby Tamarack mine had five vertical shafts ranging from 3,200 to 4,600 feet deep. Calumet and Hecla's deepest inclined shaft extended to 6,900 feet in 1902, while Quincy's deepest in 1904 was 5,280 feet deep on the incline. These were more than twice the depths achieved in Butte, Montana, and were the deepest shafts in the world. All the Michigan mines reached even greater depths in the 1910s. Quincy, for example, operated three shafts in 1920 extending roughly 6,400 feet, 7,200 feet, and 7,800 feet on the incline.[73]

The Michigan mines experienced a classic case of the law of diminishing returns as they reached ever greater depths. The lodes or veins narrowed, yielding less rock; the veins flattened out, so gravity was no longer enough to move rock from the stopes into the tramcars; and the rock became harder. To make matters worse, the rock contained less copper (see table 3.3).

New producers—like the Baltic, Champion, Mohawk, and Trimountain—were an important exception to this rule. Their deposits were not particularly rich in the early years of operating, but they improved with depth. The yield of ingot copper for the four ranged between 16 and 26 pounds per ton in 1905, about the same as for the

TABLE 3.3
Copper Yields from Rock Treated at Stamp Mills, Select Mines, 1890, 1905, and 1916

	1890[a]	1905[a]	1916[a]
C & H	64.4	48.4	30.0
Osceola	28.8	18.8	15.3
Quincy	48.8	16.6	17.5
Tamarack	49.7	21.1	18.2

Source: B. S. Butler and W. S. Burbank, *The Copper Deposits of Michigan*, U.S. Geological Survey Professional Paper No. 144 (Washington, D.C.: USGPO, 1929), 79–97.
[a]Pounds of ingot copper per ton of rock treated.

older producers delineated above. By 1916, however, their yields ranged between 21 and 36 pounds per ton, considerably higher than their older rivals.[74]

Much of the new technology put into place in the Michigan district after 1890 served partially to counteract the increased costs of working poorer deposits at greater depths. The rock tonnage hauled to the surface increased enormously, which offset to some degree the declining copper content of the rock. Quincy, for example, treated 165,000 tons of rock at its stamp mills in 1890 but 1,383,000 tons in 1911. Through the 1880s, Lake mines brought rock to the surface along single-track skip-roads in skips with a capacity of 3 to 4 tons. Starting around 1890, the major producers, including Quincy, double-tracked the skip-roads down the shafts so that they could use balanced hoisting, that is, running two skips in opposite directions simultaneously. Quincy increased the capacity of its skips to 6 tons in 1894, to 8 tons in 1902, and finally to 10 tons in 1920. More powerful engines allowed hoisting speeds to increase from about 500 feet per minute in the late 1880s to about 3,000 feet per minute by the 1910s, increasing capacity accordingly.[75]

Other technologies changed as a result of the greater working depths and the larger volume of rock handled. The man-engine, introduced in the region in the mid-1860s, was no longer an effective way to move men underground, at least at the larger mines. With the workings often several thousand feet from the man-engine shaft, men would have to waste a good deal of time getting from the man engine to their place of work. In the early 1890s, Calumet and Hecla, Quincy, and the other large producers began using specialized man-cars or cages, which ran either in separate shafts or on the skip-tracks in the production shafts.[76]

Several of the larger, more progressive mines mechanized tramming in an effort to cut costs and increase the volume of rock moved. In 1893 Calumet and Hecla experimented with a system using compressed air to power tramcars but never adopted the technology. By World War I it was using a system of rope haulage instead. Between 1901 and 1903, Quincy introduced electric haulage using electric locomotives drawing power from a trolley wire. Electric tramming saved Quincy about 8 cents per ton of rock handled, a substantial amount considering the volume of rock handled in 1905, about 1.2 million tons. Most of the remaining mines in the district, however, did not use electric haulage until World War I, when tramming locomotives powered by storage batteries became popular.[77]

The use of electricity for other applications spread slowly starting in the 1890s. Calumet and Hecla and the Tamarack mine used electric pumps underground as early as 1893, while Quincy used electric lights on the surface and in 1900 installed electric motors in new blacksmith and machine shops. Significant improvements in stamp mill recovery techniques began around 1905 with improved grinding machinery, mainly the Hardinge Grinder. From 1915 on, major breakthroughs in methods for recovering copper from sands (ammonia leaching) and from slimes (froth flotation) proceeded so quickly that by 1920 Calumet and Hecla was able to recover over 90 percent of the copper left in the sands of Torch Lake over the previous half century. During the four-year period 1917–20, C & H extracted nearly 21,000 tons of copper from the Torch Lake sands at a cost of about 11 cents per pound and at a time when copper sold for about 24 cents a pound.[78]

Taken together, the new technologies enabled the Michigan producers to partially counteract the effects of operating at greater depths and exploiting leaner deposits. They were able to reduce or at least stabilize costs and remain roughly competitive with the copper mines of Butte and the emerging producers of Arizona and Utah. In 1908 the costs of mining, concentrating, and smelting at the Michigan mines ranged between an extraordinary 7.3 cents per pound of metallic copper for the Wolverine mine and 12.6 cents per pound at the Trimountain, with most of the major producers' costs being between 9 and 11 cents. Even after transportation and marketing costs of about a half cent per pound are added to the cost, the Michigan producers still earned profits in 1908, when copper sold at 14 cents. The Arizona and Utah mines, including several open-pit operations, had costs ranging from 8.9 cents to 13.5 cents per pound of copper. For the years 1905–7, the Anaconda Mining Company's total costs may have averaged as high as 12 cents per pound.[79]

The last significant new technology introduced into the Michigan mines was the one-man drill. The Rand rock drill commonly in use in the early twentieth century weighed nearly 300 pounds by itself, but with its mounting post and clamps the complete outfit weighed 602 pounds and required two men to set up and operate. In 1909, practical lightweight drills requiring higher air pressures came onto the market, including a 90–pound model invented by J. George Leyner but developed by the Ingersoll Rand Company. In 1911 and 1912, Calumet and Hecla, Quincy, and the other mines tried several types of lightweight drills, including jackhammers as light as 30 pounds. A single man could move and operate all these new models.[80]

The mining companies enjoyed the benefits of the one-man drill as early as the spring of 1914. At Quincy, driving new drifts had cost between roughly $6.00 and $6.60 per linear foot between 1908 and February 1914, but as early as April 1914, with the one-man drill in use, costs fell to $4.80 per foot. Calumet and Hecla found that the one-man drill increased worker productivity in drifting by 91 percent and in stoping by a more modest 31 percent.[81]

The long-term impact of the one-man drill becomes clearer when we contrast operating costs in 1912, the last year the mine operators used two-man drills exclusively, with those in 1916, the first "normal" year with one-man machines in use throughout the district. Quincy, for example, produced more copper in 1916 than in 1912 but with 320 fewer miners (363 versus 683) and 347 fewer men total (1,646 versus 1,993). Miners' wages increased 43 percent between

1912 and 1916, but their productivity more than doubled over the same years. At Calumet and Hecla, stope and drift miners tripled their productivity between 1912 and 1916, and labor costs per ton of rock excavated fell by about 50 percent despite wage increases of 30 percent.[82]

The mine operators enjoyed windfall profits from the savings wrought by the one-man drill, combined with sharp increases in wartime prices until September 1917, when price controls began. Dividends paid out by the Michigan companies nearly doubled, from $12.2 million in 1915 to $23.9 million in 1917. In 1916–17, Calumet and Hecla, Quincy, and other mines produced copper at less than 50 percent of its selling price, generating huge profits.[83]

The impact of the one-man drill and wartime conditions, however, only masked for a brief time the underlying problem the Michigan producers faced: that of rising costs as the mines went deeper and the deposits became poorer. From 1901 to 1912, before the one-man drill had any significant effect, the Lake mines produced copper at costs roughly 10 percent above those incurred at the Montana and Arizona mines. This cost disadvantage increased to 16 percent in 1916 and then remained at between 25 and 30 percent over the years 1917–20, an untenable position once wartime copper prices collapsed.[84]

CHAPTER 4

The Richest Hill on Earth

The Anaconda is much too big a thing to be shut down altogether by half a dozen fires or as many quarrelsome railroads. One of the marvels of its organization is the small amount of labor it employs compared with the amount of work it performs. . . . Rock drills and black powder are the real miners. All the manual labor needed is merely to direct them—hence the paradox of high wages and cheap mining which has puzzled so many wise men of the east.—Correspondent to the *London Financial Times,* quoted in James A. MacKnight, *The Mines of Montana* (1892).

The role of mining in the development of the American West has fascinated historians for generations. Most of the published work on western mining, however, has focused on precious metals. But base-metal mining was far more important in terms of the value of product, employment, and population growth. As late as 1880, the value of precious metal production in the United States was nearly three times the value of copper, lead, and zinc output combined ($71 million versus $26 million). By the late 1890s, however, base-metal mining had pulled even with the precious metals, and by 1920 it had nearly a three-to-one advantage ($371 million versus $127 million).[1]

The extraordinary growth of western copper starting in the 1880s quickly made the West the dominant copper-producing region in the United States. The emergence of western copper mining after 1880, particularly in Montana and Arizona, is evident in table 4.1. Western copper accounted for most of the growth in the U.S. copper industry between 1880 and 1920, and in turn helped make it the largest in the

TABLE 4.1
U.S. Copper Production, by State, 1880–1920, as Percent of Total Production

	1880	1885	1890	1900	1910	1920
Alaska	—	—	—	—	—	5
Arizona	5	14	14	19	28	46
California	1	—	—	5	4	—
Michigan	84	44	38	24	20	13
Montana	2	41	42	42	26	15
Nevada	—	—	—	—	6	5
Utah	—	—	—	3	12	9
Other West	—	—	1	1	2	6
Southern states	3	—	5	1	2	1
Other	5	1	—	5	—	—

Sources: Raphael Pumpelly, *Report on the Mining Industries of the United States*, U.S. *Tenth Census, 1880*, vol. 15, pt. 2 (Washington, D.C.: USGPO, 1886), 798–800; *Mineral Industry, 1892*, 108–9; *Mineral Industry, 1901*, 176; *Mineral Industry, 1911*, 149; and *Mineral Industry, 1921*, 151.

world. From the 1910s on, the exploitation of most of the world's copper resources was controlled by the large American copper companies that began their operations in the American West. The growing importance of the American copper industry in world markets can be seen in table 4.2.

The history of the Montana copper mining industry, centered in Butte, is the focus of this chapter, while the rest of the western copper mines, mainly in Arizona, New Mexico, Utah, California, and Alaska, are considered in subsequent chapters.

Butte and the Rise of Montana Copper

Butte, known as the "richest hill on earth" because of the copper it held, began as an important silver-producing district. Butte's first substantial mining boom focused on silver but also produced discoveries of copper. On the last day of 1874, William L. Farlin relocated his

TABLE 4.2
World and U.S. Copper Production, 1880–1920

	World[a]	U.S.[a]	U.S. Share
1880	172,000	30,200	17%
1885	253,000	83,000	33
1890	302,000	130,000	43
1900	545,000	303,000	56
1910	961,000	544,000	57
1918	1,579,000	955,000	61
1920	1,057,000	612,000	58

Source: William B. Gates Jr., *Michigan Copper and Boston Dollars: An Economic History of the Michigan Copper Mining Industry* (Cambridge, Mass.: Harvard University Press, 1951), 198–99.
[a]Tons of 2,000 pounds.

Asteroid silver claim and renamed it the Travona. More than a dozen additional silver mines opened between 1875 and 1877 and many of these evolved into successful copper ventures. The development of the Montana mines followed a common pattern. The original small-time prospectors who discovered the lodes typically lacked sufficient capital to develop the properties, so they would sell their claims to mining entrepreneurs who had the capital resources. Peterson's *Bonanza Kings*, a detailed study of fifty important western mining entrepreneurs who followed this path to great wealth, includes ten men involved in Montana mining.[2]

One of the early investors in silver and copper mining was Andrew Jackson Davis (1819–1890), who came to Montana in 1864 and formed a partnership with Samuel T. Hauser (1833–1914), owner of the First National Bank of Helena. Davis bought the Parrot mine in 1877 for $10,000 and formed a partnership with two Connecticut brass manufacturers to develop the property. An inventory of tools and equipment at the Parrot, dated 23 July 1877, ended with the prediction, "One good copper mine—Parrot by name—which will, when worked, *prove of great value*."[3]

The second entrepreneur to play a major role in Butte mining, William Andrews Clark (1839–1925), also became an important player in the early 1870s. He bought four major claims, which later became the Original, Colusa, Mountain Chief, and Gambetta mines, all in 1872. Through loans to Bill Farlin, Clark gained control of his silver properties, including the Travona mine. Clark also founded the Colorado and Montana Company, which opened a smelter south of Butte in August 1879 to treat the ores from his silver mines. The smelting firm later bought three valuable copper properties in Butte—the Gagnon, the Nettie, and the Burlington. Clark had a knack for earning large profits from every venture he entered, to the point where people began referring to "Clark's luck." From his Colusa mine, for example, a claim only 175 feet long, he extracted over 30 million pounds of copper in two years.[4] The third major figure, Marcus Daly (1841–1900) came to Butte in 1876 and quickly began speculating in silver and copper mines. Daly had wide experience in western mining. He prospected for gold at Grass Valley, California, in the late 1850s; worked at the Comstock Lode in Virginia City, Nevada from 1862 to 1868; managed the Walker brothers' Emma mine in Alta, Utah, in the early 1870s; and ran their Lion Hill properties south of Salt Lake City from 1872 to 1876. The Walkers sent Daly to Butte in August 1876 to buy a silver mine, and he quickly bought the Alice mine for $25,000 and built a stamp mill to process its silver ore.[5]

The Anaconda mine, which earned its fame from copper, also began as a silver prospect. Michael Hickey, a local Butte prospector, located a silver claim in 1875 that he named the Anaconda. While serving in the Civil War, Hickey had read a column in the *New York Tribune* in which Horace Greeley predicted that Grant's forces would surround Lee's army and crush it like a giant anaconda. Hickey liked the image and used the name of the big snake for his claim. He lacked the capital to develop the property, so in the autumn of 1880 he sold it for $30,000 to Marcus Daly, who had just received that sum for his shares in the Alice mine. To develop the property, Daly in turn enlisted the financial support of a San Francisco investment syndicate consisting of James Ben Ali Haggin, Lloyd Tevis, and George Hearst.[6]

Haggin and Tevis were brothers-in-law who had moved to California during the gold rush and established a law partnership in Sacramento in 1850. They invested in railroads, telegraph lines, express companies, and irrigation projects, all in the West. George Hearst had joined the California gold rush in 1850 but made his first fortune at the Ophir mine on the Comstock in 1859 and 1860. Hearst was a self-taught geologist and prospector who joined Haggin and Tevis in 1870 and brought them mining expertise. His wealth allowed his more famous son, William Randolph Hearst, to launch a publishing career. The Haggin-Tevis-Hearst mining syndicate developed some of the most valuable metal mines of the West, including the Ontario silver mine in Utah; the Homestake gold mine in the Black Hills of South Dakota; and the Anaconda, one of America's richest copper mines.[7]

In the spring of 1881, the syndicate bought the Anaconda from Daly for $30,000, gave him a quarter of the shares, and made him superintendent. He began shaft-sinking and excavation work in June 1881 and soon produced silver at a profit. A year later, Daly's men reached the 300–foot level and the silver ore disappeared, but they discovered instead the first segments of the richest copper sulphide deposit known to man. In May 1883 they reached a depth of 600 feet, where the vein was 50 to 100 feet wide, with ore containing an average of 12 percent copper and with pockets that assayed at 50 percent. Daly then persuaded the syndicate that the Anaconda had enormous potential, but as a copper mine rather than as a silver prospect. By the late summer of 1884, the syndicate had sunk $4 million into the mine, including all the profits earned up to that point. The profitable development of the Anaconda would involve a huge investment in the mine proper but would also require large and expensive reduction and smelting works.[8]

Daly did not consider Butte as a potential smelter site because of the high price of land and the lack of water there. Instead, he built his smelter twenty-six miles west of Butte on Warm Springs Creek and named the new town Anaconda. The plant went into operation in October 1884 and initially consisted of a concentrating mill, which treated about 500 tons of ore daily, and a smelter with twenty-four roasting furnaces and twenty-six matte furnaces. Initially, the Anaconda smelter produced matte containing 64 percent copper, which then went to eastern and British refineries for further treatment.[9]

Several other mines opened in Butte and vicinity starting in the late 1870s, so Anaconda faced plenty of local competition. W. A. Clark developed a mine in the late 1870s that he named Clark's Colusa to distinguish it from the Montana Copper Company's Colusa mine. In 1884 he built a separate plant, known as Clark's Colusa Smelter, to treat his ore. Clark in turn sold the smelter to the Boston and Montana Consolidated Copper and Silver Mining Company in 1888, and it remained in operation as Boston and Montana's "Lower Works" until 1893, when the firm opened its new plant at Great Falls.[10]

The Parrot mine remained undeveloped in the late 1870s. Hauser and Holter—along with their Connecticut investors Franklin Farrel, Archille Migron, and others—founded the Parrot Silver and Copper Company on 16 August 1880, with Farrel serving as president and Migron as company agent. The firm built a smelter, which by 1883 was producing 5,000 tons of matte containing 3,300 tons of metallic copper per year. Although Hauser and Holter together owned about a quarter of the shares in the Parrot, Connecticut stockholders owned the rest.[11]

Adolph and Leonard Lewisohn, two New York investors, bought the East and West Colusa claims in Butte in 1878. The Colusa Smelter went into operation in 1880, became part of the Boston and Montana Company in 1887, and functioned as its "Upper Works" until 1893. The Lewisohns built the Butte Reduction Works (also called the Colusa Parrot) in 1883 as a custom smelter, but it had an unsuccessful debut. Nevertheless, W. A. Clark bought the property in 1887, enlarged it several times, and ran it until 1910, when a major fire destroyed the plant.[12]

Most eastern capitalists who invested in Michigan copper mines showed only modest interest in Montana copper properties. One major exception was the tandem of Boston investors Albert S. Bigelow and Joseph W. Clark, who as partners had made fortunes in Michigan copper with the Osceola and Tamarack mines. In April 1887, Bigelow warned John Daniell, the superintendent of his Michigan operations, to prepare for a secret trip to Montana to examine copper properties Bigelow wanted to buy. Daniell spent from May 12 to 28 in Montana looking at mines, including the Colusa. Following negotiations with Leonard Lewisohn, in early July Bigelow concluded the deals that created the Boston and Montana Consolidated

Copper and Silver Mining Company by combining the Mountain View mine with the Montana Copper Company. On the last day of 1887, the Boston and Montana bought several properties from William A. Clark for $150,000, including Clark's Colusa Smelter and a collection of mines: Clark's Colusa, Liquidator, Modoc Extension, Gambetta, Piccolo, and the Mountain Chief.[13]

The Boston and Montana initially controlled two smelters in Butte but soon built a concentrator, smelter, and electrolytic refinery at the Great Falls of the Missouri River, about 130 miles northeast of Butte. The Boston and Montana Company signed an agreement with the Great Falls Water Power and Townsite Company on 12 September 1889 providing for the development of the waterpower site. The Townsite Company agreed to build "a complete and durable dam across the Missouri River above Black Eagle Falls," which would supply the smelter with at least 1,000 horsepower by 1 September 1890 and at least 5,000 horsepower by 1 January 1891. The Boston and Montana agreed to build a smelter at the site costing at least $300,000. The agreement obligated both parties to minimize environmental damage from water and air pollution. The Boston and Montana was a highly successful copper enterprise throughout the 1890s, and among the Montana producers it ranked second only to Anaconda in production and profits.[14]

Charles H. Palmer, who came to Montana in 1888 from the Michigan copper district, helped launch another major producer, the Butte and Boston Mining Company. The major stockholders were Boston investors with significant interests in Michigan copper mines, including Albert S. Bigelow and Joseph W. Clark. With Palmer as manager, the Butte and Boston developed its main properties in Butte—the Mountain Chief, Silver Bow, Grey Cliff, and Matte mines—and began smelting operations in September 1888. Production in 1891 was more than 9,000 tons of metallic copper.[15]

The Butte and Boston Mining Company struggled to earn profits during its early years. Palmer was unable to generate enough output to cover operating expenses, creating serious short-term drains on the firm's working capital. A group of Boston investors holding the Butte and Boston's bonds forced the firm into bankruptcy in early 1896. The court sold the property at auction in February 1897 to a group of investors who formed the Butte and Boston Consolidated Mining Company, with a capital stock of $2 million. The new firm had virtually the same directors as the Boston and Montana, including Albert S. Bigelow, Leonard Lewisohn, Charles Van Brunt, and Thomas Nelson. Frank Klepetko, the Boston and Montana superintendent, became the manager of the Butte and Boston properties as well.[16]

During its first decade, the Anaconda Gold and Silver Mining Company remained a closed venture, entirely owned by Haggin, Tevis, Hearst, and Daly. They formally incorporated the Anaconda Mining Company in January 1891, with a capital stock of $12.5 million, which soon doubled to $25 million. The firm reincorporated on 15 June 1895 with a capital stock of $30 million as the Anaconda Copper Mining Company, the name it kept until 1955. In October 1895 the Exploration Company, Ltd., a London investment syndicate controlled by the Rothschilds, bought 300,000 shares from the George Hearst estate at $25 per share. The same group of investors bought the remaining 270,000 Hearst shares in June 1896 at $27 a share, giving them nearly half the stock, but Haggin still owned a majority interest.[17]

In an effort to cut costs and become more self-sufficient, Anaconda built its own railroad linking the Butte mines with the Anaconda reduction and smelting works. The Montana Union Railway, built by the Union Pacific and Northern Pacific railroads, connected Butte and Anaconda in 1884, but Daly feuded with the Montana Union over freight rates, switching charges, and service. This increasingly bitter business relationship persuaded Daly to build his own line.[18]

In September 1892, Daly incorporated the Butte, Anaconda & Pacific Railroad, with capital stock of $1 million entirely owned by Anaconda. Daly used an army of unemployed miners to build the line, which went into operation on 1 January 1894. The new line immediately cut freight rates in half, to 25 cents per ton of ore, and still made handsome profits. For the year ending 30 June 1898, when the B., A. & P. had a total of nearly $4.7 million in invested capital, the line earned profits of $489,000 despite charging only 30 cents per ton to ship copper ore, coal, and coke.[19]

Anaconda also struggled to improve the speed and reliability of its deliveries of matte, converter bars, and cathodes to its major customers on the eastern seaboard. Throughout the 1890s, the Anaconda Reduction Works did not have the capacity to smelt, refine, and convert all of the Butte ore into ingot copper. The Baltimore Copper Smelting & Rolling Company had contracts with Anaconda beginning in January 1891 to refine Anaconda's matte and converter bars into metallic copper. Anaconda agreed to supply the equivalent of at least 15,000 tons of copper per year in the early 1890s and 30,000 tons starting in 1896. For much of 1895 and 1896, the two firms feuded over discrepancies between the weights of the various forms of copper recorded at Anaconda and those recorded in Baltimore, with the Anaconda weights often as much as 3 percent higher.[20]

Anaconda generated large profits throughout the 1880s and 1890s. Once the Anaconda Reduction Works went into full operation, output jumped from only 1,943 tons of metallic copper in 1884 to 20,231 tons the following year. Production averaged more than 30,000 tons per year in the late 1880s, between 40,000 and 50,000 tons in the early 1890s, and surpassed 60,000 tons per year for 1896–98. For the three years ending 30 June 1898, profits from all operations were $12.9 million, or $4.3 million per year on stock with a nominal value of $30 million. Anaconda's subsidiary departments—which produced coal, lumber, bricks, water and electric service, and various other goods and services—yielded profits of $1.9 million over the same three-year period, or nearly one-sixth of the total.[21]

Although Anaconda faced stiff competition from other Butte producers in the 1880s and 1890s, it remained the dominant force in Montana, as table 4.3 illustrates. Anaconda produced half of Montana's copper until the late 1890s. The Montana Ore Purchasing Company emerged from the rest of the pack into a distant third place by the turn of the century. Not all of its production represented output from its own mine because the firm also did custom smelting for other Butte mines.

Following the first wave of concentrator and smelter construction in Butte in the 1880s, the milling, smelting, and refining of Butte's ores were increasingly centered in the giant Anaconda works west of Butte and in the Boston and Montana plant at Great Falls. Five Butte smelters closed between 1893 and 1910, leaving only the Pittsburgh

TABLE 4.3
Montana Copper Output, by Firm, 1885–1900

	1885[a]	1890[a]	1895[a]	1900[a]
Anaconda	18,000	32,000	49,890	55,400
Boston & Montana	3,750	13,471	30,373	41,600
Butte & Boston	—	2,743	—	—
Butte Reduction Works	1,250	1,500	1,700	6,228
Clark's Colusa	5,000	—	—	—
Colorado S & M	600	1,160	3,875	5,730
Montana Ore Purchasing	—	—	7,430	11,129
Parrot	4,900	4,500	3,630	5,200
Other Mines	400	100	375	1,900
Total	33,900	55,474	97,273	127,187

Sources: Mineral Industry, 1892, 110; and *Mineral Industry, 1900,* 193.

Note: Clark's Colusa was part of the Boston & Montana total beginning in 1889. The Butte & Boston first appeared in 1889, but beginning in 1894 it is included in the figures of the Boston & Montana.

[a]Tons of 2,000 pounds.

and Montana Company smelter (built in 1902) operating in East Butte until 1930, when it shut down permanently.[22]

Technology and the Costs of Production

Mining and haulage technology at the Anaconda and the other Butte mines in the late nineteenth century was similar to those in use at the Michigan copper mines in the 1880s. The Butte mines largely bypassed hand drilling, instead using pneumatic rock drills from the start. The Anaconda mine had 138 power drills in operation by the early 1890s, and 250 by the turn of the century. Hand drilling, however, lingered on in the smaller mines and in some surprising locations. As late as 1896, the Parrot mine, a midsized but profitable copper and silver operation, was still using the older technology.[23]

The Montana mines never built man-engines to move men underground and used cages rather than man-cars for that task. They also employed timber and, starting in 1898, steel "gallows-frame" head-frames rather than enclosed rockhouses. Because Butte Hill had more than three dozen major mines situated close together, many had connecting drifts. A large producer might operate only a single surface plant to serve a half dozen mines. Unlike the Michigan producers, the Butte mines used horses and mules extensively for underground tramming through World War I. Electric trolley locomotives did not finally replace four-legged haulage until the early 1920s, considerably later than the changeover in Michigan.[24]

Several other technologies differentiated Butte's mines from those in Michigan. While the latter used rows of individual timbers to support the hanging wall, Butte used the "square-set" timbering system first developed in 1860 in Virginia City, Nevada, on the Comstock Lode. The Anaconda / Amalgamated Mining Company operated a large framing mill at Rocker, three miles west of Butte, where machines cut the timbers into framework.[25]

In the early twentieth century, the Butte mines converted many of their steam hoists to operate with compressed air produced at a central plant by electrically driven compressors. The completion of the Canyon Ferry hydroelectric plant on the Missouri River in 1902, producing cheap electric power for Butte and Anaconda, encouraged the extensive electrification of all operations in the mines and smelters. The ACM Company began converting its hoists to operate on compressed air in 1912, and a year later it had already converted ten main hoisting engines and a similar number of smaller hoists. A central compressor plant supplied each of the twenty-two separate shafts with power for hoists, rock drills, and other tools. Electricity was increasingly used for water pumping, ventilation, underground and surface lighting, surface tramming, and electrolytic refining. The ACM Company electrified the Butte, Anaconda & Pacific Railroad in 1913 and began to use electrically driven hoists in the 1920s. Of the Anaconda mine's 33 hoists in service in 1925, steam powered 10, compressed air 17, and electricity the remaining 6.[26]

Smelting and refining technologies were distinct from those used on the native copper of Lake Superior. Montana smelters originally roasted raw ore, which contained about 12 to 15 percent copper, either in the open air or in stalls, to drive out most of the sulphur.

Starting at the Anaconda Reduction Works in 1883, smelter operators roasted the ore in hand-rabbled reverberatory furnaces, and from 1887 on, in Bruckner cylindrical furnaces. Further concentration took place in reverberatory matting furnaces, producing matte, usually about 65 percent copper. Wood served as the fuel for both roasting and matting until 1883, when Anaconda and the other producers began converting to coal.

Molten matte in turn went into a Bessemer converter, with air blown through it, resulting in metal that was about 99 percent pure copper. Once electrolytic refining came into use, this process resulted in 100 percent pure copper as the final product.[27]

The smelting processes used at Butte were destructive to the environment. The sulfide ore found there, chalcocite (Cu_2S) required an initial treatment using open roasting to reduce the sulphur content before smelting in a blast furnace. The smelters intermixed layers of ore with layers of logs, which would burn slowly for several weeks, giving off sulphur, arsenic oxides, and a variety of fluorides and particulates. Roasting of the ores took place either in large heaps, some as large as a city block, or in stalls, sometimes equipped with short stacks, but the damage to the environment was the same. The Montana producers used these roasting techniques at Butte but discontinued their use at the new smelters at Anaconda and Great Falls.[28]

Several Butte physicians argued that smelter smoke served as a disinfectant and thus prevented disease, while W. A. Clark claimed that the arsenic in the air gave the women of Butte attractive complexions. The Butte City Council nevertheless passed a "smoke ordinance" in December 1890 prohibiting heap roasting and mandating stacks at least seventy-five feet tall for stall roasting. The smelters ended heap roasting in 1891, but adding stacks to the stalls did little to improve the air. Only the closing of the Butte smelters between 1893 and 1910 solved the problem.[29]

The Parrot Silver and Copper Company introduced new smelting technology at its works in 1884 and 1885. Two of the company's stockholders, Franklin Farrel and Achelle Migron, owned the U.S. patent rights for Pierre Manhes' process for converting matte to metallic copper using a Bessemer furnace. Farrel built an experimental Bessemer plant there in 1884, and in July 1885 the Parrot stockholders agreed to pay Farrel and Migron royalties for the use of the Manhes process. In addition, Farrel received $25,000 in cash and 600,000 newly issued shares of Parrot stock, with a face value of $10 per share. Parrot's stockholders, however, stopped payments in June 1887 and challenged the validity of the patents. The viability of the process, however, was never in doubt. The Parrot works converted a matte of 72 percent copper into metal containing 98.8 percent copper in twenty minutes in a single operation.[30]

The development of electrolytic refining facilities in the 1890s strengthened the competitive position of the Butte mines, particularly compared with the Lake producers. Electrolytic refining resulted in a final product as pure as Lake copper while improving the recovery of the gold and silver found in Montana's ores. After an experimental electrolytic refinery began working in Phoenixville, Pennsylvania, in the late 1870s, the new process spread quickly. Edward Balbach built the first commercial electrolytic refinery in the United States at Newark, New Jersey, in 1883. The Baltimore Smelting and Rolling Company opened the next major plant in 1888 in Baltimore, followed by a rash of new refineries: the Boston & Montana plant at Great Falls (1893), the Anaconda Mining Company works at Anaconda (1894), the Calumet and Hecla Mining Company Buffalo Works (1895), the Guggenheim Smelting Company plant at Perth Amboy, New Jersey (1895), the Raritan Copper Works (Lewisohn Brothers) in Perth Amboy (1899), the Nichols Chemical Company plant in Brooklyn, and the De Lamar Refining Works in Carteret, New Jersey (1902). The refining of domestic copper by the electrolytic process, which in 1893 amounted to about 38,000 tons—roughly one-fifth of American output—climbed to 347,000 tons by 1906, or 75 percent of the total.[31]

By the late 1890s, the Butte producers were more than competitive with the Michigan mines. The Montana mines were not only operating closer to the surface, finding richer ores, and converting the ores to metallic copper more cheaply than the Lake producers, but they also enjoyed added revenues from gold and silver recovered in the electrolytic refining process. In the mid-1890s, when Michigan mines were already at depths of 4,000 feet and below, measured on the incline, the deepest Butte shaft was 1,300 feet, and only eight of the

TABLE 4.4

Production Costs of Selected Copper Mining Companies and the Copper Content of Their Ores, 1897

	Production Costs[a]	Copper Content
Montana		
Anaconda	7.56¢	4.60%
Boston and Montana	3.40	8.63
Michigan		
Atlantic	10.01	0.65
Calumet and Hecla	NA	3.08
Franklin	10.71	1.10
Osceola	11.32	1.07
Quincy	6.91	1.56
Tamarack	9.18	1.63
Wolverine	8.47	1.36

Sources: Mineral Industry, 1897, 212–13; *Mineral Industry, 1900,* 190–191; and Anaconda Copper Mining Company Papers, Montana Historical Society, box 167, folder 8.

[a]Costs per pound of metallic copper.

twenty-seven operating shafts extended 1,000 feet or more. At the turn of the century, when thirty-five mines were at work in the Butte district, only fifteen had shafts extending more than 1,500 feet, with the deepest shafts extending between 2,000 and 2,200 feet below the surface.[32] The relative costs of the Michigan mines compared with those of Butte can be seen in table 4.4.

Montana copper producers also enjoyed considerable revenue from gold and silver sales. For the year ending 30 June 1897, for example, one quarter of Anaconda's revenues of $17.3 million came from precious metals. In 1903, when Lake copper yielded 20 ounces of silver per ton of copper and no gold, Anaconda's copper generated 100 ounces of silver and 2 ounces of gold per ton of copper, and the Boston and Montana's Great Falls refinery yielded 70 ounces of silver and 1 ounce of gold.[33]

The combined effects of richer ores, more easily worked mines, and economies from large-scale operations gave Montana's producers a comparative advantage over their Michigan counterparts. With

TABLE 4.5
Output of Metallic Copper per Man in Michigan and Montana, 1890–1909

	Michigan			Montana		
	Output[a]	Labor Force	Output Per Man[a]	Output[a]	Labor Force	Output Per Man[a]
1890	50,700	6,724	7.54	55,474	1,958	28.33
1902	85,097	14,306	5.95	133,250	6,698	19.89
1909	113,624	19,575	5.80	156,919	16,697	9.40

Sources: Mineral Industry, 1892, 110; *Mineral Industry, 1900,* 193; *Mineral Industry, 1910,* 149; and U.S. Bureau of the Census, *Report on the Mineral Industries in the United States at the Eleventh Census: 1890* (Washington, D.C.: USGPO, 1892), 156–160; U.S. Bureau of the Census, Special Reports: Mines and Quarries, 1902 (Washington, D.C.: USGPO, 1905), 237, 248–49, 504–5; U.S. Bureau of the Census, *Thirteenth Census of the United States, Taken in the Year 1910,* vol. 10: *Mines and Quarries* (Washington, D.C.: USGPO, 1913), 101, 109–111

[a]Tons of 2,000 pounds.

productivity much higher in Montana than in Michigan, Montana producers could pay their miners premium wages and still earn healthy profits. These advantages continued well into the early twentieth century (see table 4.5).

The Wars of the Copper Kings and the Peace that Followed

The great size and profitability of Butte's mines helped produce a few wealthy and powerful "copper kings," who often clashed in both the economic and political arenas. The first and most important "war" involved William Andrews Clark and Marcus Daly, extended from 1888 until Daly's death in 1900, and became increasingly vicious over time. The conflict first manifested itself in November 1888, when Clark ran for the office of territorial delegate to Congress as a Democrat in heavily Democratic Montana and should have won easily, but his Republican opponent, Thomas H. Carter, beat Clark by more than 5,000 votes. Democratic Butte and Anaconda had voted

Republican. Clark quickly discovered that Daly had pressured his employees and others to vote for Carter, ensuring Clark's defeat.[34]

The origin of the Clark-Daly feud is unclear. In 1951, Kenneth Ross Toole argued that Daly backed the Republican because he was certain that there would be a Republican administration in Washington and Daly feared federal action against his timber operations in Montana. More recently, David Emmons has offered an intriguing explanation that the conflict derived from ethnic-religious hatred. Clark was the perfect Orangeman, a Scotch-Irish Presbyterian with his roots in Ulster, while Daly was an Irish Catholic. The Republican candidate whom Daly supported in 1888 happened to be an Irish Catholic.[35]

Clark might have become one of Montana's first U.S. senators in 1889, when the state legislature sent four men, including the Democrat Clark, to the Senate, two from each party. The Senate seated the two Republicans. Clark then fell three votes short of getting elected to the Senate in 1893, largely because Daly lobbied against him. Clark won the Senate seat in early 1899 through extensive bribery of state legislators, purportedly paying up to $10,000 per vote. Clark never got a chance to savor his victory, because the Senate investigated charges of election bribery brought by Daly and in April 1900 forced Clark to resign his Senate seat. Later in the year, the Clark forces took control of the Montana Democratic Party, and in mid-January 1901 the state legislature elected him to the U.S. Senate. He served without distinction for his full term (1901–1907) and in 1910 sold his remaining copper properties in Montana to the Amalgamated Copper Company for $10 million.[36]

The Amalgamated Copper Company dominated the Montana copper industry in the early twentieth century. The brainchild of Thomas W. Lawson (an unscrupulous Boston stockbroker, financier, and speculator), the Amalgamated resulted from the efforts of two Standard Oil magnates, Henry H. Rogers and William Rockefeller, to gain control of the American copper industry. Lawson's original plan was to consolidate the large Boston-based producers, including Calumet and Hecla, Quincy, and the Bigelow-Lewisohn properties (Boston & Montana, Butte & Boston, Tamarack, and Osceola). Lawson owned substantial chucks of stock in Boston & Montana and Butte & Boston, so he would have profited greatly from such a combination. But a chance meeting between Henry H. Rogers and Marcus Daly of Anaconda, combined with the difficulty of buying the closely held stock of Calumet and Hecla and Quincy, shifted Rogers's focus to the Anaconda.[37]

In late April 1899, financial journals carried news of the formation of the Amalgamated Copper Company, with capital stock of $75 million and Standard Oil chieftains Henry Rogers and William Rockefeller among its directors. The Amalgamated was a typical holding company of that era, established to control other corporations by buying and holding their stock. The Amalgamated organizers made a fortune through the initial sale of stock because the company's assets, the stock it owned, were worth only $39 million. The Standard Oil investors, who risked only $13 million of their own funds, ended up with 500,000 shares of Amalgamated, worth about $50 million. This was the "crime of the Amalgamated" Thomas Lawson referred to in his sour-grapes expose of the Amalgamated a few years later.[38]

Rogers induced stockholders in several large Montana mining companies to sell him a controlling interest. Anaconda was the largest prize. The Haggin-Tevis-Hearst-Daly syndicate owned all the Anaconda stock until 1895–96, when the Hearst estate sold 570,000 shares, nearly half the total, to the Exploration Company, Ltd., for $15 million. At the time the Amalgamated came into existence, Tevis no longer had stock in Anaconda, and the elderly Haggin sold off his interest in 1901 for $15 million. Daly, who became the first president of the Amalgamated, converted his Anaconda stock, worth about $17 million, into Amalgamated shares.[39]

Daly struggled to retain his power and influence within the newly formed Amalgamated while his health began to deteriorate. Daly suffered from Bright's disease, which nearly killed him in October 1899. He returned to work, but his health worsened and he died on 12 November 1900 in New York City. Daly spent the last part of his life fighting an effort by the Boston and Montana interests within the Amalgamated to shift all reduction and smelting work to Great Falls and to close the Anaconda plant. By 1900 the Rogers-Rockefeller-Lawson team had already gained control of the Boston & Montana/

Butte & Boston twins but did not absorb them into the Amalgamated until the following year.[40]

Daly's strategy was to increase production sharply at the Anaconda works while artificially inflating Anaconda's profits by siphoning funds from Anaconda's subsidiaries. Daly had little success in increasing production, but he improved the apparent profits of Anaconda by adeptly manipulating the company's accounts, kept in a bank that Daly and his cronies controlled. In the end, the key leaders of the Amalgamated decided to put Frank Klepetko, the Great Falls manager, in charge of the Anaconda Reduction Works with the intent of replacing the original plant rather than closing it.[41]

The Amalgamated's leaders—Rogers, Ryan, and Rockefeller—found their dealings with Clark, Daly, and Holter easy and predictable. But this was not the case when they dealt with the third major player on the Montana copper scene: F. (Fritz) Augustus Heinze (1869–1914). "Fritz" Heinze and his two brothers, Otto Jr. and Arthur, used their family inheritance and loans to launch the last major new producer in Butte, the Montana Ore Purchasing Company, in March 1893. Capitalized at $2.5 million, the M.O.P. Company built a technically advanced smelter in 1893 and did low-cost custom smelting for the smaller producers of Butte. Initially, the firm owned no mines but profitably smelted ore from leased properties and from the independents. Fritz Heinze quickly established a reputation as a shrewd judge of mining properties, an unscrupulous financial manipulator, and a master in using the courts for his financial gain.[42]

Fritz Heinze fought the Bigelow / Lewisohn interests and then the giant Amalgamated Copper Company between 1898 and 1906, and was successful enough to force the Amalgamated to buy him out. He had powerful allies in his battles. His brother Otto provided credit in amounts up to $500,000 when needed. W. A. Clark also became a Heinze ally, and Clark's banks provided credit as well. Fritz Heinze's other brother, Arthur, supplied legal expertise. Heinze also controlled two of the three members of the Second Montana District Court (Butte), which heard most of the cases involving his properties, despite the best efforts of his enemies to bribe them.[43]

Heinze's legal battles with the Boston & Montana, and ultimately with the Amalgamated, derived from the principle of "extralateral rights" to minerals, which originated in California and Nevada in the 1850s and 1860s, and which was incorporated into the federal General Mining Law of 1872. This legal provision of the 1872 law granted the owner of a claim the rights to "all veins, lodes, and ledges" that either outcropped or had their top or apex on the claim, *over their entire length,* no matter where they extended. Because the location of the "apex" of a mineral vein became the legal determinant of ownership, the extralateral rights provision of the General Mining Law of 1872 is usually called "the law of the apex." This provision resulted in expensive, time-consuming, and disruptive litigation in western mining districts, particularly at Butte. Prospectors and mine owners who discovered mineral deposits usually tried to buy up all the adjacent properties simply to protect themselves against lawsuits.[44]

In 1893, Heinze bought the Glengarry mine and discovered a rich vein of ore. He used his profits to buy the Rarus and Johnstown mines in 1895 for $400,000, and these properties in turn generated even greater wealth. Heinze also bought the Nipper, which proved to be an extension of the great Anaconda Lode. He briefly held a one-sixth share of the Michael Davitt, which he bought for $15,000, before stampeding the Boston & Montana management into buying the property for $600,000, making his share worth a tidy $100,000. The Michael Davitt soon became the focus of a complex series of lawsuits involving Heinze and the Boston & Montana.[45]

Heinze's battles began in late 1896, when Bigelow's firm claimed that Heinze's Rarus was mining veins that apexed on their Michael Davitt mine and sought an injunction to prevent Heinze from working the Rarus. In March 1898 the case went to federal court in Butte, where a jury ignored the judge's directed verdict for Bigelow and instead ruled in favor of Heinze. In 1900 the court restrained Heinze's Montana Ore Purchasing Company from mining the Rarus, but Heinze formed a new firm, the Johnstown Mining Company, which was not a party to the suit, and continued mining under that ruse. Through a combination of tactics, including ignoring court orders, Heinze mined the Rarus from 1896 to 1904 with few interruptions.[46]

Between 1897 and 1906, Heinze was able to manipulate, frustrate, paralyze, and otherwise exasperate the Boston & Montana Company and after 1899 the Amalgamated Copper Mining Company. He had a remarkable knack for finding ways to enrich himself and give his larger rivals fits. After a careful study of the claim maps of Butte, he discovered a tiny triangular plot, forty square yards in all, lying between the Anaconda mine and the St. Lawrence, previously unclaimed. In the spring of 1899 he filed a proper claim for the land, called his new discovery the Copper Trust Mine, and briefly forced the adjoining mines to close, claiming that their deposits apexed on the Copper Trust. Two years later the Montana Supreme Court dismissed the case.[47]

Heinze used a variety of tactics to frustrate his enemies, including stockholder suits, dummy corporations, and an adept use of the apex laws and the courts. He employed as many as thirty-five attorneys to handle the legal aspects of his mining ventures. The Amalgamated, having tired of the costly litigation with Heinze, finally bought out his Butte properties for $12 million in 1906. With his departure, the state courts dismissed eighty lawsuits involving claims of over $120 million, while the federal courts dismissed twenty-three additional cases. After leaving Butte, Heinze suffered from declining health and eventually lost most of his fortune.[48]

Despite its struggles with Fritz Heinze, the Amalgamated expanded and went through several mutations but eventually became Anaconda again in a much more powerful and grasping form. In July 1900 the *Miners' Magazine,* published by the Western Federation of Miners, remarked that the Amalgamated "holds the people of the state in its grip as firmly as a slimy reptile holds an innocent lamb in its coils before devouring it."[49] This observation was even more true for the enlarged Anaconda Mining Company that reappeared in 1915. Before the Amalgamated bought out Heinze in 1906, it already accounted for 85 percent of Montana's copper production and employed nearly 7,000 men in Silver Bow County alone. Following the purchase of Heinze's mines and W. A. Clark's properties in 1910, the Amalgamated enjoyed a near-monopoly position in Butte. The last of the interminable apex lawsuits ended in 1913.[50]

Between the death of Marcus Daly in 1900 and the transformation of the Amalgamated into the larger, fully integrated Anaconda Copper Mining Company in 1915, two corporate leaders emerged who would direct the firm into the mid-1950s: John D. Ryan (1864–1933) and Cornelius F. Kelly (1875–1957). Daly's immediate successor was William Scallon, who served as president of the Anaconda Copper Mining Company and directed all of the Amalgamated's Montana operations from 1900 to 1904, before resigning to return to his private law practice.[51]

John D. Ryan moved to Butte in 1901, bought an interest in the Daly Bank & Trust Company, and before the end of the year Daly's estate had appointed him president of the bank. After Ryan aided the Amalgamated in its political offensive against Heinze in 1903 and 1904, H. H. Rogers put him in charge of the Amalgamated's Montana operations in 1904 and appointed him president of Anaconda in 1905. As Rogers' health declined, Ryan spent more time at the Amalgamated's New York office and became the corporate president in 1909. Benjamin B. Thayer then served as president of the Anaconda Copper Mining Company from 1909 to 1915. Following the dissolution of the Amalgamated, Ryan served as president of the enlarged Anaconda Copper Mining Company from 1915 to 1918 and as chairman of the board from 1919 to 1933.[52]

Cornelius ("Con") Kelley was the son of Jeremiah Kelley, a longtime friend of Marcus Daly. He joined the Amalgamated legal department in Butte in 1901 and became secretary of the Anaconda in 1905 and chief counsel in 1908. Kelley became vice president of the Anaconda Copper Mining Company in 1911, succeeded Ryan as president in 1918, held that post until 1940, and then served as chairman of the board until 1955. The death of Rogers in 1909 and the buyout of Clark's remaining copper mines in 1910 marked the end of the era of the first copper kings—Daly, Clark, Heinze, and Rogers. If we view Marcus Daly and his handpicked successors—Scallon, Ryan, and Kelley—as a single administration, the Anaconda Copper Mining Company had management continuity between 1881 and 1955, a remarkable record.[53]

Early fears that the Amalgamated might monopolize U.S. copper production and, in concert with the Rothschilds and others, control the world copper market, proved groundless. Had the leaders of the Amalgamated been able to include the major Michigan mines and the Montana producers in a Copper Trust, they might have controlled American and world markets. At its peak in 1900, Amalgamated accounted for about one-third of U.S. output (one-tenth of world production) but still did not have a large enough share of the world market to dictate output levels and prices. Amalgamated tried to restrict output and raise prices between 1899 and 1901, but when the Lake mines refused to cooperate, the firm abandoned the effort. The opening of new deposits in Arizona, Utah, and Nevada after the turn of the century made any further efforts to create a cartel.[54]

When it failed to restrict copper output and control prices, the Amalgamated turned to rationalizing production at its Montana properties. The Amalgamated enlarged and upgraded the smelting and refining complex at Anaconda shortly after it acquired the Anaconda Copper Mining Company. P. A. O'Farrell, a critic of Marcus Daly, writing in 1899, argued that the original reduction plant at Anaconda, opened in 1884 (the Upper Works) and 1889 (the Lower Works), had outdated and inefficient technology for Montana ores right from the start. Using the type of steam stamp that had worked well on Michigan copper rock, for example, resulted in the loss of 35 percent of the copper values in the concentration phase. O'Farrell calculated that the outmoded plant cost Anaconda and its stockholders at least $5 million per year in lost profits. Daly and Haggin shared the blame because neither had had the foresight to make the needed changes. Anaconda's own figures confirm the magnitude of the losses of copper at the works.[55]

To be fair, Daly did not oppose new technology at the Anaconda Reduction Works. The original equipment at the Upper Works, consisted of twenty-four hand-rabbled reverberatory roasters and twenty-six matting furnaces, each with hearths measuring 18 × 12 feet. In 1890, Daly enlarged the matting furnaces to 27 × 15 feet and replaced the roasters with forty Bruckner cylinder roasters. At the Lower Works, he added fifty new Bruckners to the original equipment in 1890 and enlarged the original matting furnaces in 1898. Daly built a Bessemer converter plant with five square converters in 1892, and a second one with eight round converters in 1894. The much-maligned Anaconda founder operated an experimental electrolytic plant there between 1891 and 1893, the third of this type in the United States, before opening a much larger permanent electrolytic plant in 1894. Still, the physical layout and location of the Upper and Lower Works at Anaconda prevented Daly from upgrading the technology more rationally. The solution, which he set into motion before he died, was to replace the Anaconda Reduction Works in its entirety.[56]

The Amalgamated hired Frank Klepetko from the Boston & Montana smelter at Great Falls to design and then operate a new smelting works at Anaconda. The Washoe Copper Company, incorporated in 1899 as a separate firm with a capital stock of $20 million, built the plant and leased it to the Amalgamated. The Washoe Reduction Works, which went into operation in 1902, initially had a capacity of 5,000 tons of ore daily, compared to the combined capacity of the original Upper and Lower Works of only 4,000 tons. The Amalgamated quickly demolished the old works and increased the capacity of the new facility to 15,000 tons per day. The new plant was an unqualified success. Anaconda treated 1.6 million tons of ore there in 1905 at a cost nearly 39 percent lower than with the old works, for a total savings of $2.58 million. In addition, the new facility reprocessed the slag from the old works and earned Anaconda an additional $5.4 million after expenses.[57]

From the day the new reduction works opened, farmers in Deer Lodge Valley complained of damage to their crops and livestock. After a company study concluded that many of these claims were legitimate, Anaconda paid out $330,000 to settle all claims against it. To reduce pollution and the threat of additional lawsuits, in 1903 the Washoe Reduction Works installed a system of dust chambers in its various departments and connected them via a 2,300–foot flue to a new 300–foot stack, which rose to a height of 1,100 feet above Deer Lodge Valley. Nevertheless, the farmers initiated a series of lawsuits in 1905 to collect additional damages of more than $1 million or force the permanent closure of the smelter. Following several long trials and appeals, the U.S. Supreme Court dismissed the case in April 1909, ruling in favor of the Anaconda company on all issues. In 1919, Anaconda added a 585–foot stack, nearly twice the height of the previous structure. To this day, it is the tallest brick structure in the world.[58]

A series of improvements to the Washoe Reduction Works at Anaconda completed in 1915 greatly enhanced the recovery of copper from ore and from tailings produced since 1884. The introduction of the froth flotation process improved the recovery rate of copper from the concentrates from 82 percent to 96 percent. A new plant to treat the slimes from flotation resulted in the recovery of an additional 5,500 tons of copper annually. In addition, the Anaconda Mining Company built a leaching plant to treat the past accumulation of course tailings, about 20 million tons, which contained about 11 pounds of copper per ton. The leaching plant, which could treat 2,000 tons daily, recovered 80 percent of the copper in the tailings.[59]

The individual properties owned by the Amalgamated—including the Anaconda, Boston & Montana, Butte & Boston, and the Parrot—had remained separate and distinct operating entities. The evolution of the Amalgamated Copper Company back into the Anaconda Copper Mining Company began in 1910, when the Anaconda stockholders authorized an increase in the capital stock from $30 million to $150 million in order to buy the other copper companies operating in Butte and vicinity. The rationale was twofold: a single corporation would both reduce costly litigation and realize operating economies. In 1910, Anaconda also bought W. A. Clark's remaining copper properties in Butte—the Stewart, the Colusa-Parrot mine, the Original mine, and the Butte Reduction Works—giving the giant firm control of most of Butte's copper.[60]

In 1914, Anaconda acquired the International Smelting and Refining Company for $15 million, extending its operations well beyond Montana. As a result of the acquisition, the firm acquired copper mines and a smelter at Miami, Arizona; a lead smelter in Tooele, Utah; refineries in East Chicago, Indiana (lead), and Perth Amboy, New Jersey (copper); and a controlling interest in the Greene Cananea Copper Company in northern Mexico. Finally, on 7 June 1915, the Amalgamated shareholders voted to dissolve the company but granted themselves the right to a one-to-one exchange of their shares for Anaconda stock.[61]

As shown in table 4.6, The rationalization of production enabled the Montana copper industry to continue to expand production and remain competitive, even in the face of new deposits developed in Arizona, Utah, Alaska, and overseas. Peak production came during World War I, with 1916 as Montana's record production year for

TABLE 4.6
Montana Copper Production, 1885–1920, and Its Share of U.S. Production

	Production[a]	Share of U.S. Total
1885	33,900	41%
1890	55,474	42
1895	97,273	51
1900	127,187	42
1905	159,590	36
1910	143,121	26
1916	175,998	18
1918	161,690	17
1920	88,872	15

Sources: Mineral Industry, 1892, 110; *Mineral Industry, 1900,* 193; *Mineral Industry, 1910,* 149; *Mineral Industry, 1914,* 141; *Mineral Industry, 1921,* 151.

[a]Tons of 2,000 pounds.

copper. The state's declining share of American output reflected the emergence of new producing regions.

Management and Labor in the Montana Copper Industry

In Butte and elsewhere in the hardrock mining districts of the West, labor unions played a vital role in defining the relationship between managers and workers. Unions of western miners date from May 1863, when Comstock Lode miners founded a short-lived union in Virginia City, Nevada. The urge to organize weakened in 1864 and 1865, when the industry was in decline and miners faced unemployment and lowered wages. Following the revival of mining on the Comstock in 1866, miners established local unions at Gold Hill, Nevada, in December 1866 and in Virginia City on 4 July 1867. Over the next decade, miners created local unions in Colorado, South Dakota, Idaho, and Utah.[62]

On 10 June 1878, Butte's silver miners discovered that the mine operators had unilaterally reduced their wages from $3.50 to $3.00 per day with no advance notice. They immediately went on strike and on 13 June founded the Butte Workingmen's Union, with a constitution that mirrored that of the Virginia City miners' union. They closed the major silver mines, and through a series of strikes that extended through the end of August, the union won the restoration of the $3.50 minimum daily wage for all underground workers.[63]

The Butte Workingmen's Union, which initially included non-miners, became the Miners' Union of Butte City in 1881 and the Butte Miners' Union (BMU) in 1885. It began with 65 members in 1878 and grew to 800 by 1881 but nevertheless struggled in its early years. The group bought a building lot and began constructing a meeting hall in 1881, but in February 1882 the building collapsed. A year later, the union had only 78 paid-up members, but the rapid growth of copper mining brought in new recruits, and by 1885 the BMU had 1,800 members, the largest union of its kind in the West. The union finished its meeting hall in 1885, and its membership soared to 4,600 in 1893 and reached 6,200 by 1910, the peak of the BMU's power and influence.

The Butte Miners' Union pushed for the closed shop starting in 1885 and finally achieved that goal in June 1887. Because the Butte miners had a powerful union for about forty years and most of the city's other workers belonged to unions, Butte became the "Gibraltar of Unionism" in the West.[64]

Butte was also the birthplace of the Western Federation of Miners, among the most radical of American labor unions in the late nineteenth and early twentieth centuries. A series of violent and successful mine owners' counterattacks against local unions in Coeur d'Alene, Idaho, and in Cripple Creek and Leadville, Colorado, in 1892 prompted the larger local unions to form an alliance. More than forty delegates representing eighteen local unions in Colorado, Idaho, Montana, South Dakota, and Utah convened at the Miners' Union Hall at Butte on 15 May 1893 to form "a grand federation of underground workers throughout the western states"—the Western Federation of Miners. The history of the Butte union, which became Local No. 1 in the federation, remained intertwined with the fate of the Western Federation.[65]

Despite the BMU's occasional struggles over working conditions, underground safety, and benefits for victims of accidents and sickness, it remained a bread-and-butter union in which pay and job security were the overriding concerns. The $3.50 per day minimum wage won in 1878, combined with steady employment, made Butte's miners among the best-paid industrial workers in the United States. A six-day week was the norm, but the BMU insisted that the mine operators allow the men to work seven days a week at their option. During the 1890s, in "normal" times, they consistently earned $100 a month, roughly double the pay of industrial workers in other parts of the country. Over the period 1890–1906, the earnings of Michigan copper miners ranged between $64 and $77 per month.[66]

The rate of $3.50 per day was a minimum rate that applied to all underground workers. Miners, however, had the option of negotiating individual contracts or "leases" with management, much like the system used in Cornwall and in Michigan, which enabled some to earn twice the minimum. This system remained in place at Anaconda for only a few years after the death of Marcus Daly. Starting in 1906, when only about 10 percent of the men underground still worked on a contract basis, Anaconda replaced the teams of contract miners with gangs of men, all earning the same wage. By 1910 the old-style machine drills run by two-man teams had also disappeared, replaced by one-man machines.[67]

Compared with other western metal mining districts, the Montana copper industry enjoyed labor peace between 1878 and 1914. Anaconda faced brief strikes at its mines in 1896 and at its converter plant in Anaconda in 1898, but these episodes were exceptional. The BMU was so cooperative in its dealings with employers that it often infuriated the leadership of the more radical Western Federation of Miners. In a landmark five-year agreement with the Amalgamated in 1907, the BMU agreed to a base wage of $3.50, with higher wage rates up to $4.00 a day, depending on copper prices. In July 1912 the two parties renewed the contract for another three years. The Amalgamated signed similar agreements covering its employees at the Anaconda and Great Falls smelters from 1907 through 1915. The contract with the Anaconda Mill and Sméltermen's Union, No. 117, Western Federation of Miners, delineated pay rates for 147 different jobs within the plant, with virtually all jobs paying between $3.00 and $4.00 per day.[68]

The Butte miners were both pawns and major players in the "wars of the Copper Kings." In the early battles, the Clark-Daly feuds, they threw their support to Daly at election time. In the 1900 elections, Clark and Heinze supported the eight-hour-day proposals of the BMU in return for electoral support. Clark wanted to become a U.S. senator, and Heinze wanted to retain his control of judges whom he could use against the Amalgamated. Starting in 1902, the BMU supported the Amalgamated against Heinze and began a long period of "friendly" relations with the Amalgamated. The BMU and the giant copper producer worked together with little friction even after the Amalgamated bought the Heinze and Clark mines in 1906 and 1910.[69]

Although the BMU helped launch the Western Federation of Miners and remained in the federation until 1914, the Butte union became increasingly conservative over time. Between the Butte union's foundation in 1878 and its dissolution in 1914, it engaged in few strikes against mine operators, while the WFM led scores of violent strikes through the hardrock mining districts of the West and in Michigan. From 1902 on, the Butte local was often in conflict with the WFM, with the BMU refusing to support the federation on a wide range of issues and the more radical leaders and locals of the federation accusing the BMU of betraying union principles. A serious split within the BMU emerged by 1907, with a "progressive" faction and the old-line conservatives wrestling for control.[70]

The Butte Miners' Union was distinct from the other local unions in the WFM, reflecting the unique features of the Butte mining industry, the workforce, and the community. In *The Butte Irish,* David Emmons argues that Butte's large, stable, and conservative Irish immigrant enclave controlled the BMU and used it to promote the interests of that enclave. Butte was a large, stable copper mining community dominated by a single ethnic group. Over the period 1885–1914, the BMU had 180 officers, and 80 percent of them, including all the presidents, were Irish. Anaconda routinely closed its mines for Miners' Union Day (13 June) and for three Irish holidays: Robert Emmet's birthday, Irish Societies' Picnic Day, and St. Patrick's Day.[71]

The Butte Miners' Union had surprisingly friendly relations with the mining companies, in large part because Butte's most important capitalist, Marcus Daly, was himself an Irish immigrant. The first four presidents of the Anaconda / Amalgamated were Marcus Daly, William Scallon, John D. Ryan, and Cornelius F. Kelly, all Irish by birth or descent and all members of the Ancient Order of Hibernians. These corporate leaders spoke repeatedly about the "community of interest" shared by the company and the union. There is considerable evidence that Anaconda controlled the BMU by 1892 through a combination of tactics. The company gave its supporters within the union soft jobs at its mines but also paid them off more directly with lucrative underground contracts, gifts, and political appointments. Between about 1892 and 1912, the BMU consistently supported the Anaconda Mining Company against Clark, Heinze, the WFM, Butte socialists, and the Industrial Workers of the World (IWW). Anaconda and the BMU allowed foremen and shift bosses, who hired underground workers, to join the union as nonvoting beneficiary members.[72]

Profound divisions within the BMU came to the surface in 1912, but had roots extending back to 1907, when the union's "progressive" elements took control. The president of the BMU in early 1912 was George Curry, a member of the Socialist Party, while Louis Duncan, another Socialist, was mayor of Butte. In late March 1912, Anaconda and the other operators fired about five hundred "socialist" miners, mostly Finns. The progressive leaders of the BMU urged a strike to protest the firings, but in a referendum held on 29 March, conservatives opposed to a strike won handily, 4,460 to 1,121. The conservatives then won control of the union in June and signed a three-year contract with the Anaconda Copper Mining Company providing for a $3.50 minimum wage, with a sliding scale based on copper prices.[73]

The mine operators touched off the next major struggle within the BMU when they announced a new hiring practice, the "rustling card" system, in December 1912. Under this system, a man wanting to work first had to get a permit, or rustling card, by filling out an application form at the company office. If hired, the man surrendered the card, so if the employer fired him or laid him off and he wanted to work again, he had to repeat the entire process. The new system could be used to blacklist socialists, IWW members, or even ordinary union members. The Boston & Montana Company had used the rustling card system at its Great Falls plant continuously since 1899, but there is no evidence that the B & M used it to get rid of radicals.[74]

In December 1912 the progressives within the BMU won a referendum demanding the abolition of the rustling card and called for a strike, but the conservative union leaders did nothing. The progressives held mass protest meetings and boycotted Miners' Union Day in 1914 before disrupting the parade and sacking the Miners' Union Hall. They formed a rival Butte Mine Workers' Union on 22 June 1914, seceded from the WFM, and the next day destroyed the Miners' Union Hall with dynamite. The militant new union was ready to strike for recognition when the dynamiting of an Anaconda employment office in late August prompted the governor to dispatch the national guard to Butte and place the city under martial law. The troops arrived in Butte on 1 September, and exactly one week later Anaconda and the other operators announced the open shop. By the end of the year, the union movement among Butte's miners was by all accounts dead and buried.[75]

The union movement in Butte had a brief revival in 1917 following the tragic Speculator mine fire of 8 June 1917, when 164 men died. A spontaneous strike began at several mines on 11 June, and two days later Butte's miners established the Metal Mine Workers Union (MMWU), unaffiliated with any national union or federation. The mining companies, however, refused to meet with union representatives, who called a strike on 15 June. As the strike dragged on, war hysteria against the IWW and the MMWU grew, encouraged by Anaconda. A gang of masked men lynched IWW organizer Frank Little in Butte on 1 August 1917, and federal troops patrolled the streets starting on 10 August. Enough men had returned to work by late October to resume normal operations, and on 18 December 1917 the Metal Mine Workers' Union called off the strike.[76]

Sporadic strikes took place between 1918 and 1920, often led by the IWW, but company violence, federal intervention, and homegrown antiradical hysteria combined to defeat the workers' efforts. The severe disruption of the world copper industry following the end of the war produced sharp wage reductions and unemployment. Montana's copper industry produced about 162,000 tons in 1918 but only 89,000 tons in 1920 and then shut down for most of 1921, when production was below 25,000 tons. According to the editor of *The Mineral Industry during 1921,* "One beneficial result of the shut-down has been an adjustment of the employment situation and the elimination of the radical element from the ranks of mine laborers." The "Gibraltar of Unionism" remained an open-shop mining district from 1914 until 1934, when the International Union of Mine, Mill and Smelter Workers won a union contract from Anaconda.[77]

CHAPTER 5

The Emergence of Arizona

> In our Arizona mines, where carbonates abound, the rudest kind of furnace is sufficient to work the ores and to convert them into marketable copper. . . . The future of the Arizona copper mines is bright and full of promise, and now that avenues of transportation are open by which products can reach tidewater, we ought to be able to undersell every other copper country.
> —*Engineering and Mining Journal*, 9 July 1881

Arizona copper deposits required mining and processing technologies distinct from those found in Michigan and Montana. Native copper did not occur in significant amounts outside of Michigan, and copper sulfide deposits, such as the rich chalcocite (Cu_2S) found at Butte, occurred at only a few other locations, including Kennecott, Alaska, and Jerome, Arizona. Some deposits found in the southwestern United States included small "blankets" of chalcocite, but the deposits often occurred as veins or lenses of copper oxides or copper carbonates (compounds containing carbon and oxygen). These ores required distinct concentrating, smelting, and refining techniques regardless of the mining methods employed.[1]

The absence of cheap transportation and the presence of hostile Apaches severely restricted metal mining in Arizona and southwestern New Mexico until the late 1870s. The areas of Arizona containing the richest copper deposits were not even part of the United States until the Gadsden Purchase (ratified in 1854) added territory south of the Gila River to the United States. With the Gadsden Purchase, a group of San Francisco investors established the Arizona

Mining and Trading Company in 1854 to develop copper mines at Ajo. The firm mined the deposits until the Civil War, shipping ore of nearly 50 percent copper content to Swansea, South Wales, for smelting. Initially, they hauled ore 300 miles by mule train to San Diego, California, but later they sent ore by barge down the Colorado River to the Gulf of California and thence to Swansea. The Mexican government subsequently claimed that Ajo was still part of Mexico and sent troops to dislodge the miners.[2]

The Civil War was a setback for copper mining in Arizona. The withdrawal of federal troops increased the frequency of Apache attacks and temporarily made mining impossible. Abortive Confederate invasions of the New Mexico and Arizona territories in 1861 and 1862 only compounded the sense of insecurity and chaos. Precious metal "rushes" brought hordes of prospectors into Arizona Territory, and many of these inadvertently discovered copper deposits. Placer gold discoveries along the Colorado River in the early 1860s lured prospectors into the western part of the territory, while more permanent underground gold mines that were developed in the late 1860s in central Arizona produced more permanent settlements.[3]

By the late 1870s, copper entrepreneurs removed the two greatest obstacles to base-metal mining: hostile Apaches and the lack of cheap transportation. General George Crook defeated the Apaches, while two railroad lines, the Southern Pacific in the southern part of the state and the parallel Atlantic and Pacific (Santa Fe) through central Arizona, were under construction. Finally, two great silver rushes in the 1870s—at Globe in south central Arizona and at Tombstone in the southeast part of the Territory—brought thousands of prospectors into these copper-rich areas. Gold and silver mining continued sporadically in Arizona well into the twentieth century, but copper was ultimately the most important metal, with the value of copper production surpassing the precious metals by 1888. Over the period 1858–1960, Arizona produced well over $8 billion in metals, with copper comprising 84 percent of the total.[4]

Except for Henry Lesinsky's Longfellow mine at Clifton, begun in 1873, Arizona copper mining ventures failed miserably during the 1870s, but the rapid development of the Copper Queen mine at Bisbee in 1880 marked the turning point for the Arizona copper industry. Production for the entire Arizona Territory, perhaps 400 tons of metallic copper in 1874, stood at 1,000 tons in 1880, but by 1882 had jumped to roughly 9,000 tons, second only to the Lake Superior district. Output increased to over 11,000 tons in 1885, but by then Arizona was a distant third in total production behind Michigan and Montana.[5]

Unlike the Michigan copper mines, strung along the Keweenaw Peninsula for about a hundred miles, or the Montana mines, located exclusively at Butte, the copper industry of Arizona was located in four major districts until the 1890s, when three additional mining fields came into production. This chapter will examine the emergence of the four oldest mining areas through 1885 in the order of their development: Clifton-Morenci, Globe, Jerome, and Bisbee.

Clifton-Morenci

The development of the Clifton-Morenci district, a mountainous area located just west of New Mexico and about 150 miles north of the Mexican border, illustrates the problems Arizona mine operators faced before the coming of the railroad. In July 1870, Jim and Bob Metcalf discovered two of the major deposits in the district, later the Longfellow and Metcalf mines, but hostile Apaches prevented any mining. They returned to the area in July 1872 to complete the work required to establish title to the claims—that is, surveying the property and locating the deposits. Other prospectors were making claims, so in August 1872 the men established the Copper Mountain Mining District and elected Joseph Yankie the first district recorder.[6]

The Metcalfs could return to the Morenci area and establish legal title to these claims only because they had won financial backing from the Lesinskys, successful merchants based in western New Mexico. Henry Lesinsky (1836–1924) was a Polish Jew who emigrated to Australia in 1854 and in 1858 came to California. Around 1869, he joined his uncle, Julius Freudenthal, who ran a wholesale provisions business in New Mexico. Based at Las Cruces, in the southwest part of the territory, they supplied grain and flour to federal troops, carried the mails, and ran a stagecoach line. In the late 1860s, Henry's brother Charles joined the firm.[7]

In 1872, Metcalf, Lesinsky, and five others made a dangerous six-day trip of about 100 miles through Apache territory on horseback to examine the Longfellow mine, and on 28 August 1872 the Metcalfs, Lesinsky, and others established the Francisco Mining Company to develop the property. Lesinsky purchased a controlling interest in the venture for $10,000 in 1872 and bought out the rest of the Metcalfs' holdings in 1874 for an additional $20,000. Henry and Charles Lesinsky, Eugene Goulding, and David Abraham then incorporated the Longfellow Copper Mining Company on 1 May 1874. They had already begun working the mine in early 1873 but struggled for the rest of the decade to make a profit.[8]

The Longfellow had rich ore but suffered from high transportation and raw material costs. Initially, selective mining of a deposit that outcropped on the surface produced black ore that Lesinsky shipped to Baltimore for smelting and that yielded 35 percent copper. The ore went by bull team to Las Cruces, then over the Santa Fe Trail to Independence, Missouri, a total of 1,200 miles, and from there to Baltimore. The same wagon train returned to the mine site with food, mining equipment, and other supplies. Various observers reported the Longfellow deposits to have a copper content ranging from 15 to 39 percent. Since miners could remove most of the ore through horizontal tunnels driven into the hillsides, the Longfellow and other Clifton mines did not need expensive hoisting equipment.[9]

One of Lesinsky's partners, Eugene Goulding, began smelting in 1873 at Clifton using a brick and adobe furnace that burned charcoal made from mesquite found nearby, with blacksmith's bellows providing the blast. By 1875, eight furnaces of this type were in use. Lesinsky later employed a German metallurgist who designed a reverberatory furnace that operated for exactly one day in January 1875 before the corrosive ore destroyed the brickwork. Louis Smadbeck, a cousin of Lesinsky's from Connecticut, experimented with copper linings on the older adobe furnaces, and in 1877 he developed copper water-jacket linings. By the late 1870s, the Longfellow smelter had three rectangular water-jacket furnaces driven by waterpowered blowers. They burned charcoal as fuel and smelted about thirty tons of ore daily.[10]

Lesinsky and his partners also explored promising deposits in the mountains above Chase Creek west of Clifton. The Metcalfs developed a collection of mines known as the Metcalf group, while Lesinsky developed the Coronado and Queen mines.

The most interesting failed mining venture of the 1870s was the Detroit Copper Mining Company of Arizona, originally incorporated in July 1872 but reincorporated in March 1873 by a group of Detroit capitalists headed by "Captain" Eber Brock Ward (1811–1875), a wealthy Detroit shipbuilder and industrialist. The firm had a nominal capital stock of $500,000, with two-thirds owned by Detroiters. Seven investors from Silver City, New Mexico, owned the remaining third, probably reflecting their contribution of mining claims to the enterprise.[11]

Ward sent a party of men to survey the property and to carry out assessment (development) work and hired a retired ship captain, Miles Joy, to direct these operations. The men established a tent city called Joy's Camp, which later became Morenci. Additional exploratory work continued until Ward's death in January 1875. William Church (1845–1901), who came to Arizona in 1880, reorganized the Detroit Copper Mining Company, resumed mining, and built a smelter three miles south of Clifton on the San Francisco River.[12]

Lesinsky struggled to make a profit at the Longfellow throughout the 1870s, and the rest of the Clifton mines languished until the Apache threat diminished and the transportation system improved. Apache raids in 1879, mainly against teamsters and ore wagons, threatened to shut down the Longfellow. There was a brief respite following the death of the Apache chief Victorio in 1880, but Geronimo's band raided the Clifton district in the spring of 1882 and killed at least twenty-four people. The federal government did not completely "pacify" the Apaches until 1886, when Geronimo surrendered.[13]

Partly to neutralize the Apaches, Lesinsky abandoned the wagon road between the mine and the smelter and instead built a "baby-gauge" (twenty-inch) rail line, the Coronado Railroad, from Chase Creek below the Longfellow to Clifton, about five miles distant. Completed in late 1879 at a cost of $75,000, the line initially used mules to pull the empty ore cars uphill. Lesinsky had the first locomotive, weighing four tons, shipped during the summer of 1880 from New York City to San Francisco via Cape Horn, then by rail to

Yuma, Arizona, and finally by wagon the remaining distance of more than 300 miles. By the end of 1883, the mines in the Clifton-Morenci district had double-tracked gravity tramroads linking them with the Coronado Railroad.[14]

Besides the problems brought by poor transportation and hostile Apaches, Lesinsky also had to recruit labor for his mines and smelter. He employed Mexican workers at the smelter from the start and then Chinese and Mexican workers at the Longfellow mine in 1879 and 1880, and perhaps earlier. Arthur Wendt, who visited the Longfellow in late 1879, reported a labor force of 150, predominantly Chinese and Mexicans. Besides board, Chinese miners earned $40 a month, Mexican miners $50, and American miners $75.[15] J. D. Emersley, correspondent to the *Engineering and Mining Journal,* reported in February 1880 that Lesinsky was employing sixty Chinese and twenty Mexicans at the mine and twenty-five additional Mexicans at the smelter. Emersley made the following comments about the Chinese workers, reflecting the prevailing racism of the time:

> As experience has long ago shown, in other quarters, that Chinese labor in mines—where there are opportunities for skulking and stealing candles and blasting-supplies— is anything but profitable. I was not surprised to find that eighty men only extracted 17 to 20 tons of ore per day, while twenty white miners could easily have broken down an equal amount. I believe the output has been largely increased of late, but it will be strange if the owners do not get tired of their Asiatic employees at no very distant date.[16]

Lesinsky also faced serious shortages of both working capital and investment capital. One stopgap solution was to pay employees, especially the Mexican workers, in scrip, or "*boletos,*" which they could use only at the company stores. Departing employees could exchange their *boletos* for Mexican silver money, but at a discount. More important, Lesinsky's company stores earned profits even when the mining operations did not, allowing the Longfellow Mining Company to remain in operation.[17]

The only surviving cost information on Lesinsky's operations from the late 1870s was the result of his attempts to get better financing for the Longfellow. Arthur F. Wendt, a mining engineer and metallurgist, visited the mine and smelter in December 1879 and prepared a detailed report on the Longfellow's operating costs and profit potential, which he completed in January 1880 for "a Firm in Baltimore to whom the Owners of the Longfellow Mines had applied for Advances." Wendt reported production of 750 tons of ore per month, with an average copper content of 22 percent, by a labor force of 112 men at the mine and 27 at the smelter and tramroad. The total cost of copper, including delivery to eastern markets, was about 11 cents per pound, while copper sold for 20 cents per pound. Wendt argued that, with an annual output of at least 7 million pounds of copper, if copper prices fell to the unlikely level of 15 cents, profits would still be 4 cents per pound, or a total of $280,000 per annum. By late 1879, the Longfellow was producing at a rate of about 2.8 million pounds per year, and by the fall of 1882, production had doubled.[18]

Lacking sufficient capital to develop the property and desiring to return to "civilization," in the spring of 1882 Lesinsky authorized Frank L. Underwood of Kansas City to sell the Longfellow Mining Company. To make his property more marketable, Lesinsky signed a "sideline" agreement with William Church of the adjoining Detroit Copper Company on 12 May 1882. The two agreed to use vertical sidelines rather than the "extralateral rights" allowed under the General Mining Law of 1872 to decide rightful ownership of copper deposits, thus avoiding potentially costly litigation.[19]

Underwood engaged several mining and smelting experts from the United States and Scotland to examine the properties in order to hasten their sale to a group of Scottish capitalists. The prospectus issued on 11 August 1882 by the new venture, the Arizona Copper Company Limited, contained the consultants' reports. The prospectus argued that the Clifton mines were among the richest in the world and only lacked a railroad link to the outside world to produce profits that would dwarf those earned at the Calumet and Hecla in Michigan and the Río Tinto in Spain.[20]

In September 1882, Scottish investors bought Lesinsky's mines (the Longfellow, Coronado, and Queen) for $1.2 million and obtained the Metcalf group from Robert Metcalf for an additional $300,000. The resulting joint-stock company, the Arizona Copper Company, had a capital stock of $5 million, with most of the preferred shares held by Edinburgh capitalists. The Scots tried without success to buy the Detroit Copper Company as well, to give them control over the entire district.[21]

The lack of rail connections to the outside remained the key hurdle for the mines in the Clifton-Morenci district. The Southern Pacific Railroad, which crossed Arizona considerably south of Clifton, reached Lordsburg, New Mexico, in the fall of 1881. Lesinsky immediately began hauling coke by wagon from Lordsburg to Clifton, some 80 miles, and the use of coke instead of charcoal increased smelting volume and lowered costs. The Arizona Copper Company then built the Arizona and New Mexico Railway, a narrow-gauge (thirty-six-inch) line connecting Clifton and Lordsburg and completed in April 1884. This gave the Clifton district rail connections with the rest of the country. The completion of the line to Lordsburg promptly lowered the cost of Welsh coke at Clifton from $50 to $37.42 per ton.[22]

The new Scottish owners of the Clifton mines invested heavily in upgrading the smelter and in improving the transportation systems, and by October 1883 the Arizona Copper Company was already $1 million in debt. High operating costs combined with major cave-ins at the Longfellow pushed the company toward bankruptcy. Amid suggestions of insider trading of stocks and threats of stockholder suits against company officials, the stockholders reorganized and restructured the old firm in July 1884 as the "new" Arizona Copper Company. The firm became an efficient producer following the reorganization but suffered from falling copper prices starting in the fall of 1885 before earning profits by the end of the decade.[23]

The Arizona Copper Company built an elaborate, highly mechanized smelter at Clifton in 1883 and 1884. It had five round water-jacket cupola furnaces. Three were of 48-inch diameter (at the tuyeres), with a daily capacity of sixty tons each, built by Fraser & Chalmers of Chicago. The other two were of 36-inch diameter and 30 tons capacity, constructed by Rankin, Brayton & Company of San Francisco. Water powered the entire plant, with three steam engines as backups. A dam across the San Francisco River created a 24-foot head, and water flowed to two large water turbines via a 3,800-foot-long ditch leading to a 4,500-foot-long redwood timber flume. The firm spent about $100,000 on grading and other plant site preparations alone. One Scotsman closely associated with the success of the Clifton smelter, James Colquhoun (1857–1954), came to Clifton in the summer of 1883 to work as a chemist for the Arizona Copper Company. He was superintendent of the smelting department in 1884 and in 1892 became general manager of the company.[24]

The other significant producer in the Clifton district, the Detroit Copper Company, remained moribund until William Church reorganized the firm in 1880. Church recognized the potential value of the mines but lacked the necessary capital to develop them. While trying to find investors in New York in January 1881, he visited the offices of Phelps, Dodge & Company, a mercantile firm with no experience in mining. There, he met with William E. Dodge, the firm's senior partner. Church offered Dodge a half interest in the Detroit Copper Company in return for an investment of $30,000, which Church needed to build a smelter.[25]

Dodge initially rejected Church's proposal but asked James Douglas (1837–1918) to investigate Church's mines during a trip to Arizona that Douglas already had planned. Phelps, Dodge invested in the Detroit Copper Company in 1882 but only after James Douglas completed thorough investigations of the firm's assets and operations. Douglas visited the mine site in May and October 1881, and prepared lengthy reports for Phelps, Dodge on the Detroit Copper Company's prospects. Douglas argued that Church could produce copper at 11 cents per pound delivered to New York City, but with improved railroad transportation this could be lowered to 9 cents a pound. Douglas further argued that an additional investment of $150,000 would enable the Detroit Copper Company to increase output to between 30,000 and 40,000 pounds per day.[26]

Church built his smelter three miles south of Clifton on the San Francisco River some seven miles from the mine and equipped the plant with two 36-inch round water-jacket furnaces driven by water-powered blowers. The smelter produced more than $700,000 of copper in 1882–83, but Apache attacks disrupted operations and caused at least four deaths. Church decided to relocate the smelter to the mine site at Morenci in 1884 to improve security and to save most of the cost of shipping ore to the smelter. The innovative Detroit Copper Company president developed a system to deliver water to the smelter by using the waterpower already generated by the company's dam across the San Francisco River three miles south of Clifton. There, a 44-inch Leffel water turbine drove a pump, which moved water through a four-inch wrought-iron pipe to the smelter site, located seven miles away at an elevation some 1,700 feet above that of the river.[27]

Church also used an innovative blast furnace design at the Morenci smelter. Besides building standard 36-inch round water-jacket furnaces, he also constructed a rectangular water-jacket furnace designed by Carl Henrich and measuring 32 inches by 72 inches, with a capacity of 60 tons of ore per day. This design resulted in savings in fuel consumption of 10 percent over the more traditional round design. In 1886, Church also built a plant to concentrate lower-grade oxide ores.

Although never as large as the Arizona Copper Company, the Detroit Copper Company was a substantial producer in the 1880s. Metallic copper output, about 1.4 million pounds in 1882, doubled to 2.8 million pounds in 1883 and 1884, and then climbed to 3.3 million pounds the following year.[28]

Globe and Jerome

The Globe district, located in southeast Arizona some 160 miles north of the Mexican border and about 100 miles west of the New Mexico boundary, was best-known as a silver mining district in the 1870s and early 1880s. Benjamin W. Regan located the Globe and Globe Ledge (copper) claims in 1874, but with a silver "rush" in effect in the area in the late 1870s, the copper deposits near Globe received little attention. The Globe Mining Company and the Old Dominion Copper Mining Company were operating several mines and a small smelter by 1881 but with little success. The Old Dominion Mining Company smelter at Bloody Tanks, six miles from Globe, had a single 30-ton water-jacket furnace but ran for only three months.[29]

In late 1881, John Williams Jr. discovered the rich deposits of the Old Globe mine. The following year, Michael H. Simpson of Boston bought the Old Dominion, Old Globe, Globe Ledge, New York, and Chicago mines, and operated them as the Old Dominion Copper Mining Company. One mile north of Globe, Simpson built a smelter with two 36-inch water-jacket furnaces, each able to smelt forty tons of ore per day. He added a third furnace in 1883, but only two normally operated. Over time, only the Old Globe and Globe Ledge deposits proved profitable, although the firm continued to operate as the Old Dominion Copper Mining Company until 1886.[30]

Several other mining ventures started working between 1880 and 1883, including the Tacoma Copper Company, the Buffalo Copper Company, and the Long Island Copper Company, but all had ceased operating by late 1883, victims of high transportation costs. Even the Old Dominion struggled during its early years despite its rich ores, which had a copper content of more than 15 percent through 1883 and about 12.5 percent in 1885. Between June 1882 and June 1884, the Old Dominion produced 5,000 tons of metallic copper, which averaged 98 percent copper content, while output in 1885 remained at 2,500 tons of metal. The lack of a direct rail connection pushed costs well above those of competing copper producers in Arizona. Coke hauled by ox or mule teams some 140 miles from the nearest Southern Pacific Railway station, at Willcox, accounted for nearly three-quarters of the cost of smelting and roughly half the total cost of producing copper. Old Dominion managers preferred coke shipped from South Wales over coke from Trinidad, Colorado, or San Anton, New Mexico, because the Welsh coke had a much lower ash content and therefore a higher carbon content.[31]

The Old Dominion struggled with high transportation and production costs, suffered from severe cash-flow problems, and was nearly $1 million in debt at the end of 1884. Arthur L. Walker, a recent graduate of the Columbia School of Mines, came to Globe in August 1883 to work as a chemist and assayer for the Old Dominion Copper Mining Company, only to discover upon his arrival that his employer might not make its next payroll. Pope, Cole, and Company, owners of the Baltimore Copper Company smelter in Baltimore, took over the Old Dominion in 1884, with George A. Pope assuming the office of president. When Pope, Cole, and Company went bankrupt in 1886, their major creditor, William Keyser, took control of the Old Dominion, reorganized the firm in 1887, and gave it financial stability.[32]

The third Arizona copper district to emerge before 1885 was located next to the town of Jerome in the Black Hills (Black Range) thirty miles northeast of Prescott. A rancher and prospector, Morris Andrew Ruffner, discovered copper outcroppings there in 1878 and located two claims, the Eureka (later called the Verde) and the Wade Hampton. In December 1880, James Douglas and Professor George Treadwell, examined the Eureka claim for Professor Benjamin Silliman Jr. and two Philadelphia investors. Douglas argued that the great distance to rail connections—180 miles—made the rich copper deposits at the Eureka almost worthless.[33]

The Eureka and Wade Hampton claims changed hands several times in the early 1880s, but none of the owners did much mining. Frederick F. Thomas bought the claims in 1882 for William B. Murray and Arizona territorial governor Frederick A. Tritle, representing a group of New York investors who then incorporated as the United Verde Copper Company in March 1883. The town of Jerome received its name from one of the original investors, Eugene M. Jerome. The new company proceeded to develop the mines, build a smelter, and construct a road connecting the mine with Ash Fork, some forty miles to the northwest on the Atlantic and Pacific (Santa Fe) Railroad.[34]

The United Verde began with great promise and many advantages. The deposits were rich and easily won, with the Eureka ore at the surface and most of the ore from the Hampton taken from a single large stope reached by a tunnel. The rich sulphuret ores, which yielded about 15 percent copper in smelting, also contained significant silver. The company began operating a 36-inch water-jacket cupola furnace (capacity 30 tons) on 1 August 1883. During the first seventeen months of operation, when bad weather reduced its operation to only 352 days, the United Verde nevertheless produced 2,700 tons of copper and 270,000 ounces of silver. The United Verde employed only seventy-five men in the mining and smelting operations and paid a dividend of $97,500 in 1884. The firm built a second cupola furnace, though it ran only one at a time, and remained in operation until December 1884, when low copper prices and high transportation costs forced its closure. It was shut down from January 1885 through late 1887 but ultimately was successful.[35]

Bisbee and the Rise of the Arizona Copper Industry

In early May 1877, a civilian scout serving with a company of the U.S. Sixth Cavalry in the Mule Mountains in southeastern Arizona discovered what he thought was a silver deposit. Later, on 14 December 1877, Hugh Jones located what became the Copper Queen mine but abandoned it after finding no silver. George Warren ended up with a one-ninth share of the Copper Queen, but sometime in 1879, in typical western fashion, turned over his share to George Atkins after betting on and losing a footrace with a horse.[36]

In late 1879 and early 1880 a group of speculators invaded the Bisbee district from nearby Tombstone, intent on buying claims. One of the adventurers was Edward Reilly, who wanted to gamble on the Copper Queen property but had no capital to pursue his hunch. Louis Zeckendorf and Albert Steinfeld, both Tucson merchants, loaned Reilly $16,000, and in early April 1880 he bought the Copper Queen claims for $15,000. Reilly then went to San Francisco, where he enlisted the aid of two firms: Williams and Bisbee, metallurgists and mine developers; and Martin and Ballard, a construction firm specializing in seawalls and railroads. John Williams and his sons (Lewis, Benjamin, and John Jr.) had considerable experience in copper smelting in South Wales, northern Michigan, and in the West. On 12 May 1880, Martin and Ballard agreed to sink $20,000 into this venture and received seventh-tenths interest in the Copper Queen mine, with Reilly and Zeckendorf keeping the rest.[37]

The Copper Queen was the first Arizona mine to use what became the standard furnace design in the West—the round water-jacket cupola furnace built entirely of iron. John Williams, originally from South Wales, came to the United States in 1858, spent the 1860s in Michigan's copper district operating a smelter, and then moved west with his sons. His eldest son, Lewis, designed water-jacket furnaces for smelting lead and copper ores in Utah starting in 1873 and perfected the design by 1878, when the Williams family established the first copper smelter at Ely, Nevada.

Built entirely of iron, with cast-iron boshes, the water-jacket furnace was practically indestructible and required no firebrick, which was prohibitively expensive in the West. A handful of manufacturers began making this type of furnace in standard sizes, which they normally shipped in one piece and then installed on the site, where they connected the furnace to the water supply and to the blowers needed for the blast. Between 1880 and 1885, Arizona copper mines built 58 furnaces, with three firms accounting for most of them: Rankin, Brayton & Company (Pacific Iron Works) of San Francisco built 31; Fraser and Chalmers of Chicago fabricated a dozen; Prescott, Scott & Company (Union Iron Works) of San Francisco made 3; and the rest came from unknown sources.[38]

The Copper Queen Mine's first furnace was a 36-inch water-jacket design costing $11,000. The Pacific Iron Works shipped it in one piece by rail from San Francisco to Benson, sixty miles northwest of Bisbee, and a mule team moved it from there to the mine site in mid-July 1880. Williams had the smelter ready to operate in mid-August, but the need to enlarge the water supply to the furnace delayed the permanent startup until mid-September. More than a hundred American miners developed the mine through a large open cut while a force of Mexican workers prepared charcoal to be mixed with Welsh coke for use in the furnace. The single furnace produced four tons of copper per day, and Williams added a second furnace in December. The Copper Queen produced 2,500 tons of copper in its first eighteen months of operation, a remarkable performance.[39]

Despite the Copper Queen's good showing, the owners were skeptical about its future and decided to sell the property while it was making a profit. They then incorporated the Copper Queen Mining Company in New York state on 2 April 1881, with a capital stock of 250,000 shares valued at $10 a share. Throughout the rest of 1881, the owners tried to sell the property to European investors for $1.1 million but found no takers.[40]

Several additional mines of note began working in the Bisbee district in the early 1880s. The Neptune Mining Company of Hartford,

Connecticut, owned about ten claims, including the Neptune; the Corbin brothers of New Britain, Connecticut, had six claims; and the Arizona Prince Copper Company of New York was working the Copper Prince mine. The last firm brought a successful apex suit against the Copper Queen Mining Company in May 1883, with the court enjoining the Copper Queen from working 600 feet of the western part of their property.[41]

In the fall of 1880, James Douglas had urged Zeckendorf and other owners of the Copper Queen to buy the adjacent Atlanta mine claim but without success. Later he succeeded in persuading William E. Dodge Jr. and D. Willis James, partners in Phelps, Dodge & Company, to buy the Atlanta, and they acquired the property on 28 June 1881 for $40,000. Douglas opted to take a one-tenth interest in the Atlanta instead of a cash fee for his services.[42]

Douglas managed the Atlanta mine for Phelps, Dodge but with discouraging results. Over the first two years of operation, the firm spent $80,000 on exploration work but found no commercial-grade ore. Phelps, Dodge nearly abandoned the claim in early 1884, when Douglas persuaded the firm to spend $15,000 more to sink one additional shaft. In July 1884, Douglas found a large, rich orebody at a depth of 210 feet, assuring the success of the mine.[43]

The Atlanta's neighbor, the Copper Queen, had enjoyed early success. Between the summer of 1880 and 1 April 1884, the Copper Queen produced 89,385 tons of ore, which yielded 17,268 tons of "black copper" (96.5 percent pure copper), an average yield of nearly 19.3 percent, and had paid its stockholders $1,225,000 in nineteen dividends. The Copper Queen, however, appeared to have exhausted its ore by the spring of 1884, and superintendent Benjamin Williams was desperately sinking shafts and driving drifts to find a new supply. In July 1884, Williams discovered an extension of the Atlanta orebody on the Copper Queen property. Rather than becoming embroiled in costly "apex law" litigation over which mine owned the orebody, the owners agreed to consolidate the properties. The result was the Copper Queen Consolidated Mining Company, established in July 1885, with the former Atlanta owners (mainly Phelps, Dodge) holding two-sevenths of the shares in the new firm and the former Copper Queen owners holding five-sevenths. The original

Copper Queen investors, Charles Martin and Edward Reilly, soon sold their shares to Phelps, Dodge, which then held nine-tenths of the stock in the new company.[44]

By 1885, after more than a decade of dealing with the Apaches, improving rail connections with the outside, and developing a smelting technology suitable for its ores, the Arizona copper industry had become the third largest in the United States. During the great speculative boom period of 1880–85, prospectors had located countless claims, and mining promoters had launched hundreds of companies, but few survived the shaking-out process. At least 39 distinct companies mined some copper during this period, but only 14 produced more than fifty tons, and only 5 of these survived over the long term.[45]

The copper mining ventures that succeeded invested huge sums of capital to develop the underground deposits, build smelters, and improve transportation facilities. Arizona's production of nearly 25,000 tons of metallic copper in 1884–85 came from more than a dozen mines, but the five largest produced 94 percent of the total—the Copper Queen Mining Company (29 percent); the Old Dominion Mining Company (24 percent); the Arizona Mining Company (21 percent); the Detroit Mining Company (12 percent) and the United Verde Mining Company (8 percent).[46]

CHAPTER 6

Arizona and the Rest of the West

I was sent to Bingham [Canyon] about the end of 1898 to take charge of some development work and make a report on the possibility of handling these low-grade copper ores at a profit. . . . I made a report to Captain De Lamar in the summer of 1899 in which I expressed the belief, based on my investigations, that there was not only a very large quantity of ore that would average 2% copper, but that it could be worked profitably on a large scale.—Daniel C. Jackling, in T. A. Rickard, ed., *Interviews with Mining Engineers* (1922)

While Butte's mines enjoyed a meteoric rise in output in the 1880s and Montana surpassed Michigan to become America's premier copper producer for the last decade of the nineteenth century, the Butte mines quickly lost out to other western mines. The principal new rivals were a half dozen major producers in Arizona, but other new mines came into production in the late nineteenth and early twentieth centuries in Alaska, Utah, Nevada, and New Mexico. The Michigan and Montana producers together accounted for more than 80 percent of U.S. copper output throughout the 1880s and still accounted for two-thirds of American production in 1900. Their combined share of U.S. output then fell to only 28 percent by 1920, when Arizona alone produced nearly half the U.S. total.[1]

The copper industry of Arizona expanded production over the thirty-five years starting in 1885 at a rapid but steady pace. By 1900, Arizona was producing 57,700 tons of metallic copper a year and stood third among the major producing regions of the United States. The industry's growth from 1885 to 1900 came entirely from the four

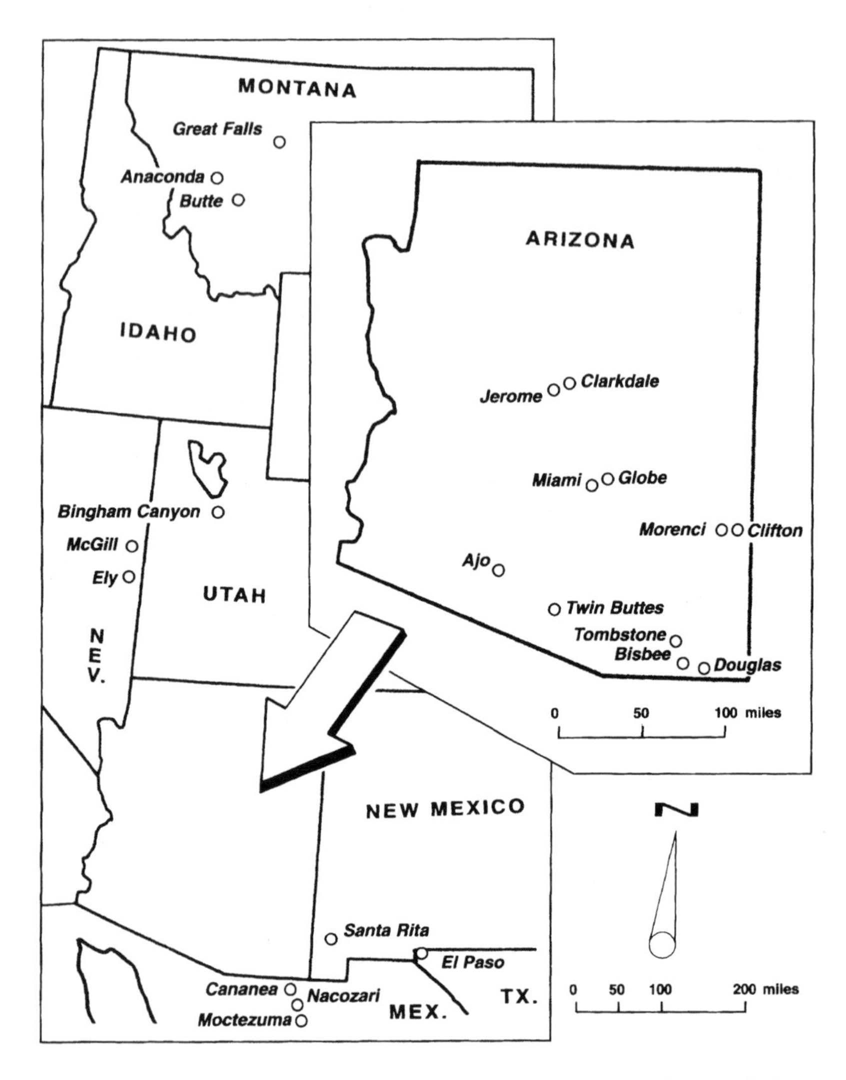

Western mining and smelting locations. (Map by Mike Brooks)

firms that had dominated production in the mid-1880s: United Verde, Copper Queen, Arizona Copper Company, and Old Dominion. The four expanded and maintained their dominance within Arizona despite drastic changes in the ores they mined. As the mines developed, the ores became progressively leaner and the carbonate and oxide ores gave way to sulphides and porphyries (low-grade sulphides), requiring new technologies for concentrating and smelting.[2]

Mining techniques changed considerably from those used in the Michigan and Montana districts. Initially the Arizona producers used traditional methods of underground hardrock mining—that is, sinking shafts, driving drifts, and stoping to exploit rich veins or orebodies at these locations. As deposits became leaner, they adopted "block-caving" mass production methods underground. Copper mining in the twentieth century has primarily involved "porphyry" copper found in highly disseminated non-vein deposits with less than 3 percent and often less than 2 percent copper. Between 1905 and 1907, Daniel C. Jackling successfully applied open-pit mining methods and innovative concentrating techniques to a vast porphyry deposit at Bingham Canyon, Utah. Since then, most of the significant new copper mining operations in the United States have used open-pit techniques.

The Arizona mine operators decided early on to negate that part of the U.S. General Mining Law of 1872 that granted "extralateral rights" to lode claims and thus avoid the costly "law of the apex" litigation that had bedeviled the Butte mines for decades. Much to the consternation of local lawyers, mine owners in some Arizona districts made informal agreements to renounce litigation. In other districts, including Clifton and Bisbee, the owners of adjoining claims signed formal legal agreements to replace the extralateral rights embodied in the 1872 law with the common-law principle of extending the sidelines vertically to establish ownership of the minerals.[3]

Arizona became the premier copper region in the United States after 1900. After jumping into second place ahead of Michigan in 1905, Arizona surpassed Montana in 1907 and 1908, fell back into second place in 1909, but was the leading producer after 1910. Arizona's output of more than 179,000 tons of metallic copper in 1920 was a fivefold increase from 1900 and constituted 46 percent of the U.S. total. A drastic shift in the makeup of the state's copper industry also took place. In 1920 the four older dominant firms—Arizona Copper, Copper Queen, Old Dominion, and United Verde—accounted for only one-third of Arizona's output. Six new giants—Calumet & Arizona, Inspiration, Miami, New Cornelia, Ray Consolidated, and

United Verde Extension—accounted for 56 percent of Arizona's copper production. Many new producers exploited low-grade porphyry deposits using open-pit mining techniques.[4]

Arizona Copper to the Forefront

The producers in the older districts—Clifton, Globe, Jerome, and Bisbee—encountered ore deposits of lower copper content, requiring new methods of concentration and smelting. The industry also faced falling prices for copper. For example, when the Arizona Copper Company came into existence in 1882, copper prices were 19 cents a pound in New York City but fell to 11 cents in 1885, below the cost of production. The Secrétan Syndicate raised world copper prices to roughly 15 cents from 1888 to 1890, but following the collapse of the copper syndicate, prices fell to 10 cents a pound for 1892–1895. Despite these difficulties, the Arizona Copper Company earned small profits and paid its first dividend in 1895, thirteen years after its optimistic birth.[5]

By the mid-1880s, the two major producers in the Clifton-Morenci district had exhausted their rich oxide ores, with 15–35 percent copper, and began mining oxide ores with only 6–8 percent copper. They could no longer smelt raw ore, and in 1886 both firms built concentration plants, the Detroit Copper Company at Morenci and the Arizona Copper Company at Clifton. William Church designed the Detroit Copper Company plant, which used Cornish rollers and jiggers to concentrate ores of 6.5 percent copper into slimes containing 23 percent copper. James Colquhoun designed a similar mill in Clifton for the Arizona Copper Company. Both plants were very wasteful of copper, leaving nearly half in the tailings.[6]

James Colquhoun became the general manager of the Arizona Copper Company in 1892 and remained in charge until 1904. His company faced a crisis in 1892, when copper fell to 10 cents a pound, barely covering the out-of-pocket costs of production. Colquhoun nevertheless persuaded company management to spend $100,000 on a plant that would recover copper from the concentrator tailings by leaching with sulfuric acid. The plant opened at the end of 1893 and was an instant success. This and various economies enabled the firm to reduce the cost of its copper at New York City from an average of 10.4 cents per pound in 1891 to 7.4 cents in 1895. Over the same period, the company was also able to double annual production, from 2,800 tons to 5,700 tons.[7]

Once the Copper Queen mine in Bisbee proved successful, Phelps, Dodge pursued other copper properties in the West, with James Douglas serving as the firm's adviser. In December 1887, Douglas persuaded the owners to buy the United Verde property in Jerome, but the deal came unraveled at the last minute. Phelps, Dodge did, however, buy out William Church's interest in the Detroit Copper Company for $1.5 million in early 1897. Walter Douglas, the son of James Douglas, worked in Morenci from April 1897 through December 1898, improving concentrator and smelter operations at the Detroit Copper Company.[8]

The changing nature of the copper deposits found in the Clifton-Morenci district required new mining and processing methods. In 1892 the Detroit Copper Company and the Arizona Copper Company discovered large deposits of low-grade (1–4 percent) sulphide copper ore in Humboldt Hill on the north side of Morenci. Both firms built sulphide concentrator plants in 1895, but only the Arizona Copper plant succeeded, by using a "wet" concentration method. By 1905, most of Arizona Copper's ore was from a series of open cuts in the side of Humboldt Hill, and plans were underway to create a single large open pit to be worked by steam shovels.[9]

The new ores also brought a change in smelting technology. The Arizona Copper Company completely rebuilt its Clifton smelter in 1901, when it installed five roasting furnaces, each with a daily capacity of 250 tons of concentrate. These produced matte copper, kept in a liquid state in reverberatory furnaces before conversion into blister copper in one of nine 7.5-ton-capacity Bessemer converters. The company later built a smelter two miles south of Clifton, where slag disposal was easier and stack emissions less of a nuisance. Mining engineer Louis D. Ricketts (1859–1940) designed the new facility, begun in 1912 and completed in August 1913. The plant was an up-to-date smelter equipped with roasting furnaces, reverberatory furnaces, Bessemer converters, and dust chambers.[10]

Besides the two long-standing producers, the Clifton-Morenci district spawned additional copper mining ventures, mostly short-lived. Only one significant new producer, the Shannon Copper Company, emerged at the turn of the century. Charlie Shannon, a nephew of Jim and Bob Metcalf, who first found copper at Clifton in 1870, discovered several claims in the early 1880s and kept the Shannon but was unable to develop it for lack of capital. He finally sold the mine in 1899 to William Boyce Thompson (1869–1930), a Montana-trained mining promoter and financier based in Boston. Thompson organized the Shannon Copper Company, which began mining in 1901. The company built a smelter on the west bank of the San Francisco River a mile south of Clifton, with a pair of 300-ton-per-day furnaces. The smelter operated from May to September 1902 without a concentrator, which Thompson built later in the year. The Shannon became the second largest producer in the district in 1909, after the Arizona Copper Company. By 1915 the quality of the Shannon's ore began to decline, and in 1919 the Arizona Copper Company acquired the property.[11]

Phelps, Dodge and Company, which had bought the Detroit Copper Company in 1897, gained complete control over the Clifton-Morenci district in 1921. The Arizona Copper Company had explored a large low-grade deposit known as the Clay Orebody and in 1917 estimated that it included at least 47 million tons of ore containing between 1.30 and 1.60 percent copper. The Clay Orebody extended into the property of the Detroit Copper Company, which held another 26 million tons. The Scottish owners of Arizona Copper lacked the capital resources to develop this deposit, and the postwar slump in copper prices forced them to stop operating entirely in May 1921. The Phelps Dodge Corporation bought Arizona Copper in October 1921 by exchanging $50 million in Phelps Dodge stock for the Arizona Copper shares. Low copper prices until the end of the 1920s, however, prevented development of these low-grade ore bodies, which ultimately yielded more than one billion tons of useable copper ore.[12]

The Old Dominion Copper Company was the only substantial producer in the Globe district between 1883 and 1911, but high transportation costs and low copper prices produced low profits and complete shutdowns in 1886–88, 1894–95, and 1897–98. Still, between 1888 and 1892 the Old Dominion produced a total of nearly 16,800 tons of metal, more than 4,000 tons per year. In 1895 William Keyser sold the property to the Lewisohn Brothers, who launched the Old Dominion Copper Smelting & Mining Company, based in Boston. The Lewisohns then hired S. A. Parnell as superintendent, replacing Arthur L. Walker, who had managed the property since 1891. The Old Dominion built its third smelter about three-quarters of a mile north of Globe in 1891 and 1892 and equipped it with three 42-inch water-jacket furnaces. Phelps, Dodge, which had bought mining properties near Globe since the 1880s, combined its holdings to form the United Globe Mines in 1893, and in 1903 purchased the Old Dominion.[13]

The Magma mine, some thirty miles west of Globe, went into production in the 1910s. William Boyce Thompson bought an inactive mine near Superior in Pinal County in 1910 for $130,000 after incorporating the Magma Copper Company. Exploration uncovered a vein of ore containing 10 percent copper 1,500 feet beneath the surface. After Thompson built a concentrator, smelter, and railroad, Magma began full-scale mining in 1915, when the firm extracted 44,000 tons of ore with 8.20 percent copper content. Production jumped from 3,000 tons of copper in 1915 to 5,500 tons in 1918 before falling off sharply after the end of the war. Magma later revived and became a major producer after World War II.[14]

In the Jerome district, the United Verde mine operated only from 1883 to 1885, when low copper prices and exhausted high-grade ore supplies forced its closing. In 1887, James Douglas, who had examined the United Verde back in 1880, persuaded William E. Dodge Jr. and D. Willis James, the principal owners of Phelps, Dodge, to buy the Unite Verde. In late December, Phelps, Dodge offered to pay off the United Verde's debts, spend an additional $200,000 on the property, and guarantee the stockholders a return based on production. The United Verde directors instead proposed an outright purchase for $300,000, which Phelps, Dodge rejected. William A. Clark, the Montana copper magnate, examined the property in March 1888 and soon bought 70 percent of the stock. By the time of Clark's death in 1925, he had purchased 95.5 percent of the stock, and by the late 1920s his family had obtained all but 1,000 shares of the total of 300,000, making the United Verde the largest privately owned copper mine in the world.[15]

Clark invested more than $2 million in developing the United Verde. He built a 26-mile narrow-gauge rail line connecting Jerome with the Santa Fe Railroad branch extending from Prescott to Phoenix. He also built a smelter with a yearly capacity of 21,000 tons of copper at the mine site in Jerome, but the settling of the ground underneath the smelter and the pollution problems it created eventually forced him to build a new plant at Clarkdale. To encourage the Santa Fe to build a branch line to his new smelter, Clark loaned the railroad $3.5 million. The railroad then built a 38-mile standard-gauge branch line, which opened in 1912. When completed in 1915, the Clarkdale smelter had a capacity of 27,000 tons of copper per year.[16]

Clark's United Verde deposit consisted of an enormous lens of sulphide ore with 7–8 percent copper and a high sulphur content. The smelter at Jerome heap-roasted the ore with cordwood for as long as six weeks, smelted it in one of six water-jacket furnaces, and then refined the metal in six Bessemer converters. Because the ore had a high sulphur content, sometimes over 50 percent, spontaneous combustion underground was a frequent problem. Efforts to stop the fires, including building concrete bulkheads, failed. In August 1902, a fire that had burned in the upper levels of the mine since 1894 broke into a working level and forced a five-month shutdown. The United Verde finally moved to an open-pit operation starting in 1918, but nearly a decade passed before the company finally brought the fires under control.[17]

The United Verde had a remarkable record of production and profits during Clark's long tenure as owner. Metallic copper output, just under 1,000 tons in 1889, reached 5,500 tons in 1894, doubled to 11,000 tons in 1896, and then doubled again to 22,000 tons in 1899. The property usually produced between 15,000 and 20,000 tons of copper annually between 1900 and 1914, with a record output of 36,000 tons in 1917. United Verde generated $10 million a year in profits in the early twentieth century, and Clark reportedly turned down a $100 million offer for the property.[18]

Following Clark's success, speculators laid claim to the lands surrounding the United Verde, but only one, the United Verde Extension Gold, Silver & Copper Mining Company (better-known as simply

U.V.X.), had the United Verde lode on its property. In 1912, James S. Douglas ("Rawhide Jimmy," the son of the James Douglas of Copper Queen fame) and George E. Turner gained control of the U.V.X., and in December 1914 they discovered a five-foot vein of 45 percent copper, the last large high-grade orebody discovered in Arizona. The U.V.X. produced 18,000 tons of copper in 1916 and more than 31,500 tons in 1917, when it was the fourth largest producer in the state. Ore taken from the U.V.X. for the entire year of 1916 averaged 23.5 percent copper, and in April 1917 it averaged 32.5 percent.[19]

Bisbee, however, remained the most important copper district in Arizona throughout the late nineteenth century, and Bisbee was dominated by the Copper Queen Consolidated Mining Company. The firm nearly fell victim to falling copper prices, which dropped from 13 cents in mid-1885 to 8 cents by early 1886. Although the Copper Queen was already in debt to the tune of $150,000, James Douglas persuaded the Phelps, Dodge partners to borrow an additional $150,000 for a new cost-efficient smelter and a railroad to connect Bisbee to the transcontinental lines. Douglas built a new smelter equipped with four 36-inch water-jacket furnaces, completed in July 1887. Shortly after that, the French Secrétan Syndicate agreed to buy 28.2 million pounds of copper from the Copper Queen Consolidated over three years starting 1 January 1888 at prices between 12.25 cents and 14.25 cents per pound. According to James Douglas, "The $300,000 of debt evaporated like dew." Between 1887 and 1906, the Copper Queen paid its stockholders $6.5 million in dividends.[20]

As the Copper Queen expanded its underground operations, Douglas revamped the Copper Queen smelter in 1893, replacing two of the 36-inch furnaces with 42-inch furnaces and installing Bessemer converters. The enlarged smelter had a capacity of about 16,000 tons of copper per year but still could not process the growing production from the Copper Queen. Phelps, Dodge also purchased mines at Nacozari in Sonora, Mexico, in 1897 and decided to build a single modern plant to handle the ores from Bisbee and Nacozari. The firm located its new smelter on a 300-acre site near the Mexican border some twenty-four miles southeast of Bisbee and named the adjacent town Douglas in honor of James Douglas. The smelter opened in late 1903 and initially had an annual capacity of 48,000 tons, quickly increased to 72,000 tons. In 1906 the Copper Queen mine alone produced almost 40,000 tons of metallic copper.[21]

The second major producer to emerge in the Bisbee district began with a claim known as the Irish Mag, named after a young woman who worked in Bisbee's tenderloin district. The Lake Superior and Western Development Company sent the brothers James and Tom Hoatson to Bisbee in 1899 to look for promising mining properties, and they bought the Irish Mag and ten other claims for $500,000 in November 1899.[22]

The Hoatsons found a rich ore mass at the 750-foot level of the Irish Mag in January 1901 and then organized the Calumet & Arizona Mining Company to absorb the Lake Superior and Western Development Company. The Hoatsons soon located a massive orebody on their property, with oxide ores as rich as 30 percent and sulphides that averaged 15 percent copper content. The Copper Queen could have claimed much of this ore, but the two firms agreed in March 1902 to waive the law of the apex and adopt the common-law practice of deciding ownership of ores at depth by extending the sidelines down vertically.[23]

Other new producers emerged in the Bisbee district in the early twentieth century, including the Shattuck-Arizona Mining Company (1904), the Denn-Arizona Copper Company (1905), and the Lake Superior and Pittsburgh Copper Company (1907). By 1911 the Calumet & Arizona owned 2,031 acres of mineral lands in the Bisbee district and was a major operator, but its output was still only about a third of the Copper Queen's. The statistics of copper production between 1880 and 1921 show the relative importance of the mines within the Bisbee district. Copper Queen led the way with nearly 900,000 tons, followed by Calumet & Arizona's 440,000 tons and Shattuck's 55,000 tons. The next largest producer, Denn-Arizona, had a total output of only 4,500 tons.[24]

Space does not allow a comprehensive discussion of all the copper mines begun in Arizona between the 1870s and 1920. Hundreds existed on paper but never mined a pound of ore. Scores of other copper ventures engaged in serious mining and smelting, sometimes for decades, without breaking even, much less earning a profit and paying a dividend to stockholders. A few of the failed mining ventures were outright frauds.[25]

The Rest of the West

The rapid growth of the Arizona mines during the first two decades of the twentieth century, combined with the large production from Butte, Montana, meant that those two states continued to dominate the U.S. copper industry through the late 1910s. Arizona and Montana combined accounted for 60 percent of American copper production in 1920, but new western mines had come into production during the first decade of this century. By 1920, mines in Alaska, Utah, New Mexico, and Nevada accounted for nearly a quarter of U.S. copper production. The bulk of this new supply of copper came from only four sources: the Kennecott Copper Corporation mine in Alaska; the Utah Copper Company mine in Bingham Canyon, Utah; the Nevada Copper Company mine at Ely, Nevada; and the Chino mine in Santa Rita, New Mexico.[26]

Although Russians and Americans had traded with Alaskan Native Americans for copper nuggets since the late 1700s, the copper resources of this vast land remained unknown until the very end of the nineteenth century. Among the swarm of prospectors coming to Alaska and northern Canada with the Klondike gold rush of 1897–98 were two veteran prospectors from Arizona—Clarence Warner and Jack Smith. In August 1900 in the Copper River Valley in the south central part of Alaska Territory, some 125 miles northwest of the port of Cordova, they found what became known as the Bonanza claims. One of the mines they developed took on the name of the nearby Kennicott Glacier, named for the Alaskan explorer Robert Kennicott. A minor typographical error in the incorporation papers gave the name "Kennecott" to the company that developed the mines and eventually became one of world's largest copper producers, the Kennecott Copper Corporation.[27]

Stephen Birch, a young New York mining engineer who had come to Alaska seeking his fortune, purchased the claims with the assistance of two well-heeled friends, Norman Schultz and H. O.

Havemeyer, who invested $262,500 for that purpose. Birch and his partners then organized the Alaska Copper and Coal Company and proceeded with development work, spending about $400,000 to buy adjoining claims. They assembled an impressive collection of properties and successfully defended their ownership rights in the courts. Their long-term goal was to attract outside capital to finance development.[28]

Birch approached the investment bank J. P. Morgan & Company with a proposal to sell the Bonanza claims. Morgan, however, was willing to invest only if the Guggenheims, experienced in metal mining and heavily involved in the giant American Smelting and Refining Company (ASARCO), would participate and manage the operations. Morgan and the Guggenheims established the Alaska Syndicate in 1906 and bought a 40 percent interest in Alaska Copper and Coal in November 1906 and the remaining 60 percent in early 1909, for a total of $2,987,500. The Morgan-Guggenheim alliance thus acquired thirty lode and eighteen placer claims encompassing about 3,000 acres. Stephen Birch stayed on as manager of the Kennecott mine, the Syndicate's most promising property. He later managed the Alaska Syndicate's central office in New York City, and in 1916 he became the first president of the Kennecott Copper Corporation. With the purchase of the adjacent Mother Lode Coalition Mines Company in 1919, Kennecott gained an additional eighty-five claims encompassing 1,500 acres and thus control over all the deposits in the area.[29]

Before the Alaska Syndicate could exploit the rich copper deposits at Kennecott, a transportation system had to be built to connect the mines with the coast. The Syndicate built the Copper River & Northwestern Railway from the new port of Cordova to Kennicott, some 195 miles away. Construction extended from 1907 until March 1911 and cost $20 million, a reflection of the extraordinary problems created by topography, permafrost, and climate. Because this railroad crossed raging rivers, sheer rock canyons, and several glaciers, half of its total length consisted of bridges and trestles.[30]

The ore initially taken from the Bonanza mines was incredibly rich chalcocite, which produced huge profits for the syndicate and later for the Kennecott Copper Corporation. The first shipment, which went by rail to Cordova and then by ship to the Guggenheim smelter in Tacoma, Washington, consisted of 1,200 tons of ore containing 70 percent copper. By the end of 1912, the mine had shipped 29,000 tons of ore containing 20,000 tons of copper, an average yield of nearly 70 percent. Kennecott produced about 32,000 tons of metallic copper in 1916 at less than 5 cents per pound, including smelting, when copper was selling for 20 cents. In 1916, workers at Kennecott produced on average of nine times the quantity of copper per man as their counterparts at the Anaconda mines in Butte, Montana.[31]

Kennecott's technology changed considerably during the 1910s. The Alaska Syndicate built a small concentrator at the Kennecott mine in 1910 and enlarged it in 1915 to treat "low-grade" ores, which contained between 7 and 20 percent copper. Given the enormous distances to smelters, Kennecott had to concentrate these "low-grade" ores to mine them profitably. The industry's standard acid leaching process was ineffective on Kennecott's carbonate ores, so in 1916 the corporation built the first ammonia leaching plant in the world at Kennecott to recover copper from its ores and tailings.[32]

Copper production in Alaska before Kennecott had been so insignificant that statisticians lumped it with other minor producing states until 1904, when it stood at 1,000 tons, a tiny part of U.S. output of 409,000 tons. Alaska's output jumped from a mere 2,500 tons in 1910 to 16,300 tons in 1912 and then to 32,300 tons in 1915. The territory averaged 50,000 tons per year in 1916–17 before falling off to 33,000 tons in 1920, a time of general depression for the American copper industry. During the 1910s and 1920s, Kennecott usually accounted for 90 to 95 percent of the total.[33]

The Porphyry Coppers

The question of who should receive the credit for pioneering the exploitation of porphyry copper is a matter of some dispute. A. B. Parsons' massive study *The Porphyry Coppers* (1933) argues that Daniel C. Jackling (1869–1956) led the way with a pilot concentrating plant at the Utah Copper Company mine at Bingham Canyon, Utah, in 1905, followed by full-scale production in 1908. Parsons admitted that James Colquhoun had previously mined low-grade ores at the Morenci mine of the Arizona Copper Company, but he argued that this did not count because the ores contained more than 3 percent copper.[34] Also in 1933, James Colquhoun published *The Story of the Birth of the Porphyry Coppers*, in which he claimed that he had a successful pilot plant for concentrating porphyry ores operating in 1896 and a full-scale plant in 1898, and that by 1901 he was exclusively processing ores with less than 3 percent copper content. The controversy is unsolvable, since much rests on the definition of the term *low-grade ore*. Although Colquhoun was an important pioneer, Jackling was more significant because he also introduced a new method for mining copper—the use of steam shovels and railroads in an open-pit setting.[35]

The work of Daniel C. Jackling at Bingham Canyon beginning in 1903 paved the way for the economic exploitation of the porphyry deposits. Enos A. Wall, who had developed mines in Colorado, Montana, and Idaho, and in other parts of Utah, came to Bingham Canyon in July 1887, and by 1900 he owned nineteen claims covering 200 acres, collectively known as Wall's Copper Property. Joseph R. De Lamar, a former ship captain, bought a one-quarter interest in Wall's Bingham Canyon holdings for $50,000 in May 1899 and had an option to buy an additional quarter interest for $250,000. De Lamar had two of his own engineers, Jackling and Robert C. Gemmell, conduct a thorough examination of the property from June through August 1899. Their report, issued in September, concluded that De Lamar could profitably mine the massive ore deposits, which assayed at 2 percent copper. They included specific recommendations for mining methods (steam shovels), the location of the concentration and smelting plants, and the appropriate technology.[36]

Following his 1899 report for De Lamar, Jackling worked as a consulting engineer in Colorado Springs for Charles M. MacNeill and Spencer Penrose, owners of the United States Reduction & Refining Company. Jackling returned to Salt Lake City and in January 1903 persuaded Wall to give him an option to buy half of Wall's Copper Property for $350,000. MacNeill and Penrose agreed to finance the purchase, and on 4 June 1903 Jackling incorporated the Utah Copper Company, with a nominal capital stock of $500,000. Following further negotiations, Jackling and his backers purchased an 80 percent interest in Wall's Copper Property for $545,000. They reincorporated in New Jersey on 22 April 1904 and increased the firm's capital to $4.5 million.[37]

Jackling started construction of a 300-ton-per-day experimental concentrating plant at Copperton in lower Bingham Canyon in June 1903. The plant went into production in April 1904, and Jackling expanded its capacity to 1,000 tons per day. He initially mined the deposits using underground tunnels and traditional stoping methods in order to allow himself time to plan open-cut operations. As work went on, the need for massive infusions of capital became evident.[38]

Jackling and his partners persuaded the Guggenheims, who controlled ASARCO, to invest in the Utah Copper Company, but they did so only after conducting a thorough assay, which cost them $150,000. In the closing months of 1905, the parties agreed that the Guggenheim Exploration Company would underwrite a $3 million issue of bonds for Utah Copper and would buy $4.64 million of its shares. The stock purchase gave Guggenheim Exploration a one-quarter interest in the mining company. In return, Utah Copper awarded ASARCO a twenty-year contract to smelt all its concentrates at a generous contract price of $6 a ton, and ASARCO built a giant smelter at Garfield, about two miles from the concentrator.[39]

Jackling moved quickly to develop the property. He built a concentrator with a daily capacity of 3,000 tons of ore, which he quickly doubled to 6,000 tons. Part of the concentrator started up in June 1907, but the entire plant did not go into operation until November 1908. The ASARCO smelter, costing $3 million, started working in August 1906 and could initially treat 300 tons of concentrates daily. Jackling hired Robert C. Gemmell to serve as general superintendent of all the Utah operations, and the two visited the Mesabi Range iron mines in Minnesota to observe steam-shovel mining operations there. They used their first steam shovel at Bingham Canyon in August 1906, and by late 1908 they had eight steam shovels and seventeen steam locomotives at work but were still producing one-third of the ore from underground operations. A year later, underground mining had virtually ended.[40]

The Utah Copper Company was not successful until it merged with the Boston Consolidated Mining Company, which owned more than 400 acres in Bingham Canyon immediately south and southwest of Utah Copper's lands. Samuel Untermeyer, a New York lawyer who was the attorney for both companies, arranged a merger in March 1910. Untermeyer increased the nominal capital of Utah Copper to $25 million, with $16.2 million in shares issued. With the merger, Jackling renamed the Utah Copper concentrator the Magna plant to distinguish it from the Arthur plant of the Boston Consolidated. By that time the Magna plant had already reached a daily capacity of 10,000 tons. In 1910–11, Utah Copper also built its own rail line, the Bingham & Garfield Railroad, a twenty-mile line connecting the mine with the concentrating plants, at a cost of $3.3 million.[41]

Jackling's success at Bingham Canyon was spectacular. During the development stage, between July 1904 and June 1907, when only the experimental concentrator was operating, the mine produced less than 2,500 tons of copper per year. In the eighteen months ending on 31 December 1908, however, Utah Copper produced 27,000 tons of metallic copper, and for 1910, an impressive 42,500 tons. During the peak wartime production years (1916–18), output averaged 95,000 tons per year. This achievement was even more impressive given the declining tenor of the ore. During the first four years of production, the ore averaged slightly above 1.90 percent copper, but during the peak wartime production years, the average copper content of the ore was only 1.33 percent. Through 1918, Utah Copper treated 80 million tons of ore averaging 1.4 percent copper and earned $142 million in profits, with $92 million paid out in dividends and the remaining $50 million retained as working capital.[42]

Their success at Bingham Canyon led the Guggenheims to invest in porphyry mines elsewhere in the West. The first was the Nevada Consolidated Mining Company mines near Ely, Nevada, in the east central part of the state near the Utah border. The town, named after the Vermont copper magnate Smith Ely, had been the scene of a gold discovery in the late 1860s. Two California miners, David Bartley and Edwin F. Gray, came into the area in the summer of 1900 and paid $3,500 for options on a number of claims thought to contain gold or silver. One of these, the Ruth claim, contained a large deposit of 2 to 4 percent copper ore. In the fall of 1902, they sold the property to Mark L. Requa for $150,000, and Requa then organized the White Pine Copper Company to develop the mine.[43]

Requa also bought a group of claims at Copper Flat, a mile west of the Ruth mine, from the Boston & Nevada Copper Company in late 1904. He incorporated the Nevada Consolidated Copper Company in November 1904, with a nominal capital of $5 million, and the new firm purchased the White Pine Copper Company and the Boston & Nevada Copper Company in January 1905. As often happened with newly launched western mining ventures, the owners of Nevada Consolidated looked for outside sources of capital.[44]

The banking firm of Hayden, Stone & Company agreed to finance the needed plant and equipment, including a railroad 150 miles long connecting Ruth with the Southern Pacific Railroad at Cobre in northern Nevada. At the same time, William Boyce Thompson purchased another group of claims near Ely and in 1905 launched the Cumberland-Ely Copper Company. The Guggenheims bought a large block of shares in Cumberland-Ely and in the spring of 1906 acquired 40 percent of the Nevada Consolidated shares for $5 million. They gained majority control in late 1906 and in August 1910 merged Nevada Consolidated with Cumberland-Ely.[45]

The two companies had already been operating as one in many respects. They became equal partners in the Steptoe Valley Milling and Smelting Company, which operated a concentrator and smelter at McGill, some fifteen miles north of Ely. The concentrator went into service in May 1908. The Nevada Consolidated was unusual among the porphyry mines controlled by the Guggenheims in that it had its own independent smelter, produced all its electricity requirements, and ran open-pit and underground mining operations simultaneously. In early 1910, at the insistence of the Guggenheims, the Utah Copper Company bought a majority interest in Nevada Consolidated stock, but Daniel Jackling did not assume direct control of Nevada Consolidated until 1915. From that point on, Jackling's staff of managers and engineers jointly directed operations at Bingham Canyon, the Nevada Consolidated mines, the Ray mine in Arizona, and the Chino mine in New Mexico.[46]

During World War I the Nevada Consolidated Mining Company became one of the largest American producers and made Nevada a leading copper-producing state. Nevada's copper production was insignificant as late as 1906 but climbed dramatically to 32,000 tons in 1910 and more than 42,000 tons in 1913 before falling off temporarily in 1914 and 1915. Wartime output averaged more than 55,000 tons per year from 1916 to 1918 before declining again at the end of the war. During the peak production years, Nevada accounted for slightly more than 5 percent of U.S. copper output. Over the decade of the 1910s, Nevada was the fifth largest copper-producing state in the United States, after Arizona, Montana, Utah, and Michigan.[47]

Investors also developed porphyry deposits in Arizona, following Jackling's lead in Utah and Nevada. Two major new mines in the Globe district, the Miami and the Inspiration, emerged in the 1910s, both exploiting low-grade porphyry deposits. J. Parke Channing, a mining engineer representing Adolph Lewisohn, visited Globe in November 1906, examined a low-grade deposit six miles west of the city, and took options on the property. After extensive exploration work, Channing concluded that the orebody contained at least 2 million tons of 3 percent ore. In March 1908 the Lewisohns organized the Miami Copper Company in New York, with a capital stock of $3 million, and began developing the property. The Miami started producing in 1911 and expanded rapidly, reaching a wartime peak output in 1918 of 29,200 tons of metal.[48]

After Channing began developing the Miami mine, William Boyce Thompson and others bought properties adjoining it, and in 1909 they founded the Inspiration Consolidated Copper Company. Thompson was a major promoter and financier of copper mines in the West, including the Shannon (1899–1901), Cumberland-Ely (1905), and Magma (1910). Thompson had the connections and charisma he needed to develop these porphyry mines.[49]

By 1912, Louis D. Ricketts was managing the development of the Inspiration mine property. Exploration and development extended over four or five years, and full-scale production did not begin until late 1915, after the owners had invested a total of more than $13 million. Ricketts decided to use an innovative process for treating porphyry ores—the "froth flotation" method for recovering copper in the concentrating plant. He built an experimental plant in 1915 that could treat 14,400 tons of ore per day, later increased to 21,000 tons. The Inspiration produced more than 10,000 tons of copper in 1915, and the following year output was a remarkable 61,300 tons, profitably extracted from ores that averaged 1.55 percent copper.

Production fell in the late 1910s, but in 1920 the Inspiration nevertheless produced 40,000 tons.[50]

The redevelopment of the Ray mine in Pinal County is more typical of Arizona copper mine development in the early twentieth century. The English company Ray Copper Mines, Ltd., which bought the Ray mine in 1898, could not profitably produce copper from the low-grade ore found on the property. Three consulting engineers examined the deposits and made widely varying assays of the ore, ranging from 1.76 percent to 5.54 percent copper. These contradictory results from legitimate mining consultants must have left the owners puzzled.[51]

Philip Wiseman and Seeley Mudd, both mining engineers, acquired options on the Ray mine and other adjacent properties and in 1906 established the Ray Consolidated Copper Company and the Gila Copper Company. Daniel C. Jackling became involved and united the two firms under the name of Ray Consolidated in 1907. Thorough but time-consuming exploration revealed that the site had at least 80 million tons of ore containing about 2.25 percent copper. The Ray Consolidated Copper Company produced 7,500 tons of copper in 1911, more than 26,000 tons in 1913, and a peak of 44,500 tons in 1917, making it the state's second largest producer.[52]

The New Cornelia Copper Company mine at Ajo was another low-grade Arizona deposit developed during World War I. In 1900, John R. Boddie, a St. Louis dry-goods salesman, bought several claims near Ajo and founded the Cornelia Copper Company, named after his wife. By 1906 two other firms were operating at Ajo, the Rendall Ore Reduction Company and the Ajo Copper Mountain Mines Company. The Cornelia Copper Company built a processing plant costing $30,000 and featuring the fanciful and fraudulent McGahan Vacuum Smelter, which was to operate without fuel and "deliver" all the elements in the ore via a series of spigots on the side of the furnace. The Rendall Company went with the equally fraudulent Rendall process, which promised to capture 98 percent of copper in ores of all types at a cost of a dollar per ton of copper. All three enterprises failed during the panic of 1907, but John Boddie reorganized the Cornelia Copper Company as the New Cornelia Copper Company in 1909. J. Parke Channing, representing the Lewisohns, took an option on the New Cornelia, but after examining the deposits briefly in late 1909 and early 1910, he decided the ore was too poor to be exploited profitably.[53]

In 1910, the mining engineer John C. Greenway came to Bisbee as the general manager for the Calumet & Arizona Mining Company. He asked Ira B. Joralemon, the Calumet & Arizona's geologist, to recommend an ore deposit that would lend itself to open-pit development. Joralemon visited the New Cornelia mines, and in the fall of 1911 the Calumet & Arizona took an option to buy 70 percent of the New Cornelia Copper Company's stock. Greenway and Joralemon hired the E. J. Longyear Company of Duluth, Minnesota, to conduct extensive testing of the orebody through diamond drilling. Longyear was doing similar work on the adjacent Ajo Consolidated Company property. Diamond drilling began in late 1911 and continued for two years. When this extensive exploration was completed in September 1913, Longyear estimated that the New Cornelia deposit included more than 40 million tons of ore with 1.51 percent copper, roughly 32.5 million tons of which was accessible by steam shovel. The Ajo Consolidated Company, which combined the former Ajo and Rendall properties in 1912, discovered reserves of 21 million tons of 1.55 percent ore as a result of diamond drilling. The New Cornelia Copper Company acquired Ajo Consolidated in 1917, giving New Cornelia total known reserves of 61 million tons of low-grade ore.[54]

Greenway faced a fundamental problem with the New Cornelia deposits: the 40-million-ton deposit consisted of roughly 28 million tons of sulphide ores situated under 12 million tons of low-grade (1.54 percent) carbonate ore. While Jackling had already shown the feasibility of mining low-grade sulphide ores profitably, there was no known process for concentrating such low-grade carbonates. Removing the overburden at New Cornelia with no return in copper would have made the development of the sulphide ores prohibitively expensive. Greenway—with the aid of Joralemon, James S. Douglas, the metallurgist Louis D. Ricketts, and a half dozen other mining engineers hired as consultants—developed an improved leaching process that used sulfuric acid to treat the carbonates. Over a three-year period extending into early 1916, Greenway and Ricketts conducted lengthy experiments in the laboratory at the Calumet & Arizona's smelter at Douglas and at Ajo, where they built experimental units with capacities of 1 ton per day and 40 tons per day. In early 1916, Greenway and Ricketts recommended that Calumet & Arizona build a 5,000-ton-per-day heap leaching plant at Ajo, which they argued would produce copper for 8.5 cents per pound, 5 cents below the market price.[55]

The New Cornelia plant went into full-scale production in May 1917, and the enterprise produced almost 10,000 tons of copper in 1917, nearly 25,000 tons in 1918, and roughly 20,000 tons per year in 1919 and 1920. This sulfuric acid leaching process recovered 75 percent of the copper contained in the ores, slightly less than the 82 percent recovery rate projected by Greenway and Ricketts. Leaching accounted for all the production at New Cornelia through 1923. Treatment of the sulphide deposits began in January 1924 in a newly finished 5,000-ton-per-day flotation concentrator. Greenway and Ricketts had confounded the prevailing views of metallurgists and mining engineers while making the New Cornelia venture a success.[56]

The last major new porphyry mine developed before 1920 was the Chino mine at Santa Rita, New Mexico, in the southwest part of the state. The first engineer to propose mass mining of the porphyries at Santa Rita was John M. Sully, an assistant superintendent for the Hermosa Copper Company, which operated a mine in Hanover, New Mexico. In December 1905 the General Electric Company, which owned Hermosa Copper, ordered Sully to examine the ore deposits at Santa Rita. He assayed 4,300 ore samples and recommended that General Electric buy the property and proceed with large-scale mining using steam shovels, but GE dismissed the plan as impractical. Sully, however, persuaded the investment bank of Hayden, Stone & Company to finance the Chino Copper Company in June 1909—but only after Daniel Jackling had examined the property and given his approval. Jackling gained control over the property, and Sully became the general manager.[57]

Jackling and his associates went ahead quickly with the development of the mine, using key engineering staff from Bingham Canyon as supervisors and consultants, particularly for the construction of a concentrator. Steam-shovel stripping began at Chino in late September 1910, and a 5,000-ton concentrator went into operation in October 1911 at Hurley, ten miles southwest of Santa Rita. Chino smelted its concentrates at the ASARCO smelter in El Paso, Texas. Once the Chino mine achieved full-scale production, it made New Mexico a major copper-producing state. Production rose from a mere 750 tons of metallic copper in 1911 to 24,350 tons in 1913 and then peaked at 55,010 tons in 1917, with the Chino mine accounting for nearly all the total.[58]

Developing the porphyry deposits of the West was a more capital-intensive, long-term proposition than had been the case with previous copper enterprises in Michigan, Montana, and Arizona. A new generation of mining engineers/promoters—Jackling, Channing, Greenway, Ricketts, and others—needed to induce investors to advance vast sums of capital to develop enormous deposits of low-grade ores. At Miami and at Bingham Canyon, mining engineers used traditional methods of exploration using shafts, drifts, and crosscuts, but these proved to be far too costly. Promoters and developers of porphyry mines increasingly conducted extensive exploration of ore bodies by using innovative "prospecting" technologies—the diamond drill for hard rock formations and the "churn drill" for softer rock. Churn drilling established the size and content of the ore bodies at Inspiration and Ray, while diamond drilling accomplished the same at Morenci and New Cornelia. Inspiration successfully floated a bond issue of $6 million in 1912, based largely on estimates produced by churn drilling that the orebody contained 45 million tons of 2 percent ore despite the fact that "no one had actually laid eyes on more than a minute fraction of that ore."[59]

Open-pit mining using steam shovels and railroads to extract and move the ore was not universally used on the porphyry copper deposits of the West, but over time it became the dominant mining technique. Jackling effectively combined the two new technologies at the Utah Copper Company mine at Bingham Canyon, and the New Cornelia (Ajo) mine started as an open-pit operation. Steam-powered shovels were first used in metal mining in 1902 at Biwabik, Minnesota, on the Mesabi Range, and Jackling first used them at Bingham Canyon in August 1906. The first generation of power shovels moved along railroad tracks and loaded ore directly into railroad cars on parallel tracks. In the early 1920s, the Utah Copper Company led the way in modifying the original technology by mounting the shovel on a set of flexible chainlike tracks that turned and thereby moved the shovel. Electric power replaced steam.[60]

An equally innovative group of mining engineers developed several new underground mining systems that allowed vast quantities of low-grade ore to be mined and transported to concentrators cheaply. At the Ohio Copper Company mine, also in Bingham Canyon, Felix MacDonald introduced an innovative undercut block-caving system of mining in 1906. The Inspiration mine used an adaptation of the Ohio system in 1911, and other underground porphyry mines developed their own variations.[61]

The rapid emergence of the porphyry mines as major low-cost producers was a remarkable development. In 1908 the Utah Copper Company had finished its first year of operations with a full-size concentrator, Nevada Consolidated had just started production, and the Morenci mine was in its second year of production, but the three mines already produced nearly 58,000 tons of copper, 12 percent of the total U.S. output. Eight porphyry mines were operating in 1918: Utah Copper, Nevada Consolidated, Morenci, Miami, Ray, Inspiration, New Cornelia (Ajo), and Chino. In only ten years, output from the porphyry mines had increased sixfold, to 343,700 tons, some 35 percent of American production. In time, all the porphyry mines used open-pit techniques, but in the late 1910s, four of the eight (Miami, Ray, Inspiration, and Morenci) still ran as underground, undercut block-caving operations. Nevada Consolidated used both methods, while Utah Copper, New Cornelia (Ajo), and Chino were exclusively open-pit operations using steam shovels and steam railroads.[62]

Management and Labor

The owners of western copper mines could operate their properties for most of their history without any significant interference from organized labor. Except for a Western Federation of Miners local at Globe, which survived with some minor interruptions between 1896 and 1917, no other local unions functioned for long at the major copper camps. This is not to say that labor-management relations were without conflict. Strife surfaced during the mid-1880s, the early years of the new century, and particularly during World War I. Significant strikes took place at Globe (1896, 1915, 1917), in the Clifton-Morenci district (1903, 1915, 1917), at Ray (1900, 1915), Bisbee (1894, 1907, 1917), and Jerome (1917).[63]

A large labor force of Mexicans and Mexican Americans were a major barrier to labor solidarity. In the twentieth century, Arizona copper workers often engaged in an unequal struggle against large, wealthy corporations owned by absentee capitalists, including Phelps Dodge, the United Verde Copper Company, the Arizona Copper Company, and the Calumet & Arizona Mining Company. Mexicans were a significant part of the workforce at the mines in the Clifton-Morenci district from the start of mining. Henry Lesinsky hired Mexican workers to operate his first smelter at Clifton in October 1873 and soon had them working in the Longfellow mine as well. At the turn of the century, when Graham County had a population of about 14,000, there were 3,096 residents born in Mexico, with most of them living in Clifton. According to Joseph Park, Mexicans comprised 74 percent of the 700 mine employees at the Detroit Copper Mining Company in 1911. In contrast, they comprised exactly one-third of the United Verde Copper Company labor force in May 1917 but only 18 percent of the employees of the Copper Queen Branch of Phelps Dodge three months later.[64]

Labor unions first appeared in the Arizona mines in April 1884 with the establishment of a union of silver miners at Tombstone, followed by copper miners' locals at Globe and Bisbee the following month, all modeled after the Comstock unions. The first well-documented case of labor unrest at an Arizona copper mine involved the issue of Mexicans working at the Old Dominion Mine in Globe. In late 1895, new management at this mine reduced the minimum wage from $3 a day to $2.75 and then to $2.25 while simultaneously hiring Mexican aliens. The Globe miners objected to the introduction of Mexicans but had no complaint about the employment of Mexican Americans. After the miners threatened the superintendent with violence, he restored their pay and promised to fire the Mexicans. When he reneged on his promises, the Globe Miners' Union called a strike, but the mine owners shut down the mine shortly after that.[65]

A major strike involving the entire Clifton-Morenci district in 1903 was the indirect result of attempts to limit the use of Mexican labor by mine operators. In January 1903, the territorial legislature enacted a law limiting underground work to eight hours a day, aimed directly at companies that hired Mexicans willing to work ten or twelve hours. The eight-hour law went into effect on 1 June 1903, and on the following day the mine operators at Clifton and vicinity announced a 10 percent pay cut to accompany the reduction of the workday from ten to eight hours. The men struck against the pay cut on 3 June and quickly shut down all the mines in the area. Most of the strikers, including the leaders, were Mexicans, with no ties to the Western Federation of Miners or other labor unions.

At the start of the strike, Governor Alexander O. Brodie ordered the entire force of Arizona Rangers (230 men) to Clifton, and the Detroit Copper Company obtained a restraining order against the strike leaders prohibiting them from pursuing the strike. A devastating flood on 9 June, which destroyed the Mexican community and killed about fifty people, badly disrupted the strike efforts. Within days, the strikers had returned to work. If nothing else, the Clifton strike must have shown the Western Federation of Miners the willingness of Mexican workers to strike but also the willingness of the mine operators to prevent unionization even at great cost.[66]

Western Federation efforts to organize Arizona copper workers following the Clifton strike illustrates the severe obstacles the union faced. By August 1905 the WFM had established nine chartered locals in Arizona, but these were fledgling organizations with little hope of success. The union's experience in Bisbee is instructive. John B. Clark, a WFM organizer from Globe Local No. 60, began signing up members in Bisbee in January 1906. In a so-called secret ballot election held in March, the Bisbee mine workers voted 2,288 to 428 against forming a miners' union. Following the election, the companies fired more than four hundred men who had voted for a union. In January 1907 the WFM sent three organizers to Bisbee, but the mining companies countered with large-scale layoffs and a pay raise. Partly out of frustration, the WFM called a strike at Bisbee on 10 April, demanding union recognition and an end to blacklisting. The strike carried well into the summer, but the companies maintained production with their regular employees and with strikebreakers. The national financial panic that began in October caused a depression in copper prices and production for the Bisbee producers. By Christmas the WFM had called off the strike, conceding defeat.[67]

The next significant labor strife began in the Globe-Miami district in January 1915. Skilled craftsmen and miners at the Inspiration Consolidated Copper Company walked out in an effort to restore wages cut in August 1914. Before the end of January, the Old Dominion, Inspiration, and Miami companies had all announced a return to the old pay scale but with one interesting twist. They adopted a sliding scale of wages based on the market price of copper, a system known throughout Arizona simply as the "Miami Scale." In early July 1915, the predominantly Mexican workforce at Ray went on strike, demanding an eight-hour day for surface workers, an end to the contract labor system, and the Miami Scale. Although the Anglo workers at Ray opposed the strike, the WFM locals at Globe and Miami rendered a good deal of assistance to the strikers. The executive board of the WFM strongly supported the Ray workers, who won most of their demands and settled the strike after only two weeks.[68]

The struggle was quite different in the Clifton-Morenci district, where the WFM initiated a strike against all the mining companies on 11 September 1915. The managers of the three principal producers—the Arizona, the Detroit, and the Shannon companies—announced that they would meet representatives of their men but only after the WFM had left the district. Although the strike remained peaceful, the managers and their families escaped to El Paso on 3 October and steadfastly refused to bargain. The companies also established a tent city at Duncan, thirty-five miles south of Clifton, as a temporary haven for about 2,000 men opposed to the strike. Governor George W. P. Hunt, who sympathized with the strikers and who had tried to mediate a settlement, called up Arizona's small National Guard for duty in Clifton, but besides a few isolated incidents, the strike remained nonviolent. Hunt ordered the soldiers to remain neutral in the strike and refused the mine owners' demand that he request federal troops.[69]

The workers effectively continued the work stoppage through the end of 1915, when increasing copper prices encouraged the owners to settle, but they did so largely on their own terms. On 11 January the strikers renounced the WFM, which withdrew its local union charters, and the men instead affiliated with the Arizona State Federation of Labor. On 24 January 1916 they voted to return to work, and marathon bargaining finally produced a new pay scale on 25 February and a contract signed on 29 March. The companies granted pay increases of 5 to 15 percent while bringing general pay scales in line with the Miami Scale. Perhaps the most important provision of the contract was the elimination of discrimination in pay between Mexicans and Anglos.[70]

The settlement of the Clifton strike of 1915–16 marked the high point of organized labor's limited success in the Arizona copper mines. The mining companies, led by John Greenway of the Calumet & Arizona and Walter Douglas of Phelps, Dodge, launched a well-organized and effective counterattack against organized labor. In 1915 they formed a state "chapter" of the American Mining Congress for the explicit purpose of lobbying the Arizona legislature to rescind legislation hostile to their interests and to forestall the passage of additional detrimental laws. Starting in 1914, Phelps, Dodge also aggressively purchased newspapers to control the news as part of a counteroffensive against organized labor.[71]

The mining companies also ousted George W. P. Hunt from the governor's chair, although only temporarily, in a disputed general election of 7 November 1916. The Republican candidate, Thomas E. Campbell, with support from the mine operators, beat Hunt by a scant 30 votes out of 58,000 cast in an election filled with irregularities. The Arizona Supreme Court overturned the election results on 22 December 1917 and awarded Hunt the governorship. A year later, Campbell defeated Hunt and again served as governor. The copper companies, however, had a less hostile, but not always pliant, governor in the person of Thomas Campbell for all of 1917, and more important, had Hunt out of the way.[72]

The Arizona mine operators encouraged the growth of the Industrial Workers of the World (IWW) in the copper camps in 1917. The companies hired detectives to pose as IWW supporters, to infiltrate the IWW, and eventually to serve as agents provocateurs. They initially intended to disrupt and destroy the established local unions of the International Union of Mine, Mill, and Smelter Workers (formerly the WFM). Once they accomplished that goal, they moved to expel the IWW and thus gain the open shop, their real goal all along. A series of strikes in June and July 1917 brought the utter defeat of organized labor in all the major copper districts. Organized labor never again enjoyed substantial political or economic power in Arizona.[73]

In case after case, mine operators who earlier had been willing to negotiate with the WFM took more severe actions following the Clifton strike of 1915–16, particularly once the IWW arrived on the scene. A strike at Ajo in November 1916 lasted two months and ended with the complete defeat of the miners. In mid-June 1917, the WFM local at the Swansea Consolidated mine at Swansea won the Miami wage scale without a strike. In late June, however, the IWW arrived and made additional demands, but the superintendent simply fired the entire workforce, brought in new men, and resumed work.[74]

Developments at Jerome illustrate the growing complexity of the labor environment in many Arizona copper camps. Three unions competed for workers in the mines of the United Verde Copper Company—the Mine, Mill, and Smelter Workers (MMSW); the IWW; and the Liga Protecta Latina, with a membership of about five hundred Mexican miners. The MMSW called a districtwide strike on 14 May, demanding an end to discrimination against unionists, a wage increase to the Miami Scale, and a closed shop. All three unions went on strike and cooperated for about two weeks. On 28 May, United Verde agreed to grant the Miami Scale and establish a grievance committee but would not agree to a closed shop. The Liga Protecta Latina returned to work on 30 May, and the MMSW accepted the terms on 3 June by a narrow majority vote.[75]

Labor unrest had exploded throughout Arizona's copper districts by early July 1917. The IWW called strikes at Jerome, Bisbee, and Miami on 26 June, and the MMSW locals were drawn into these conflicts, although the International Union refused to support the strikes. Part of the workforces at Jerome and Bisbee struck on 26 June, followed by 10,000 workers at Clifton-Morenci on 1 July and roughly 8,000 Globe-Miami workers the next day. According to one estimate, 25,000 mine workers were on strike by 6 July, virtually crippling Arizona copper production. Workers made different demands in each district, where the dimensions of the struggles and the outcomes were distinct.[76]

At Jerome, IWW officials called a second strike on 5 July to organize the remaining mines of the district and to show solidarity with their fellow workers striking at Globe and Bisbee. They also seemed intent on destroying the more conservative union already in place. The Jerome Miners' Union (MMSW) voted by a 2-to-1 margin to refuse to support this strike. With only minor violence on the picket lines, the "business leaders" of Jerome formed a vigilante force and on the morning of 10 July, rounded up about seventy-five Wobblies and deported them from Jerome by rail. The well-guarded train eventually released the deportees at Kingman, more than a hundred miles west of Jerome. The IWW nevertheless continued to attack the Miners' Union at Jerome and gained control of it a year later.[77]

The mine operators and the IWW also launched an assault against the MMSW local at Globe. On the heels of the IWW's first major organizing drive in Arizona in February and March 1917, the mine operators began discharging members of the MMSW in the Globe-Miami district in April and replacing them with nonunion men and Wobblies. After months of continuous harassment, the MMSW local voted on 30 June, by a margin of 1,437 to 416, to strike the mines in the district. The IWW also ordered its members not to work, but they were a small minority of those on strike. The Miners' Union successfully shut down all the mining operations, and with minor exceptions, Globe remained calm.[78]

After the Globe strikers roughed up a U.S. Army observer on the Fourth of July, he asked for troops to restore order. Within a few days, about 900 soldiers had come to Globe. Walter Douglas, the president of Phelps, Dodge and a major stockholder in Old Dominion, arrived in Globe on 11 July in his private railway car. He met for nearly a day with Arizona Governor Campbell, who was in Globe trying to maintain the peace. Douglas declared that no compromise with the strikers was possible and urged the community to drive the "agitators" out of town. By early August the mines were resuming production using new men hired from the outside. Although the mines did not return to full production until the end of the year, the operators had defeated the strike by mid-October.[79]

Although the workers at the Clifton-Morenci mines were nominally affiliated with the Arizona State Federation of Labor because of the settlement of the 1915–16 strike, they went on strike on 1 July

1917 to force the copper companies to abide by the "concessions" they had granted in 1915. The men remained on strike through the end of October, when they agreed to return to work and accept a retroactive decision by the President's Mediation Commission concerning grievance procedures. In time, the War Industries Board refused to allow pay increases to Mexican workers to create wage parity with Anglo workers. In short, the federal government ratified the racist practices of the Arizona copper mining companies.[80]

Labor's challenge to the copper mine operators in Bisbee began in June 1917 and ended with the infamous "Bisbee deportation" of more than 1,200 men on 12 July, which marked the complete defeat of organized labor in the Arizona copper mines. In the early months of 1917, detectives, Wobblies, and detectives disguised as Wobblies infiltrated the remnants of the old Bisbee WFM local. A dormant IWW local also came to life in January 1917, and the IWW union for copper workers, the Metal Mine Workers Industrial Union No. 800, held a statewide conference in Bisbee on 15 June. Against the advice of the two IWW organizers working in Arizona, Frank Little and Grover Perry, the Bisbee Wobblies presented a list of demands to the mine operators on 26 June. The mine managers ignored the demands, and the Wobblies went on strike the next day. The Bisbee IWW called the strike in the name of the Mine, Mill, and Smelter Workers local, forcing its International Union to revoke the local's charter and disclaim responsibility for the strike.[81]

The striking miners did nothing during the first two weeks of their struggle to justify the vigilante actions taken against them. Even Phelps Dodge's *Bisbee Daily Review* had no specific cases of violence to report. The mining companies supported the creation of two vigilante organizations before the Fourth of July—the Bisbee Workman's Loyalty League and the Bisbee Citizens' Protective League. Harry C. Wheeler, the sheriff of Cochise County, then deputized them to create a "posse" of over 2,000 men.[82]

Following a well-conceived plan that John Greenway, the manager of the Calumet & Arizona Mining Company, devised and that Walter Douglas of Phelps Dodge supported, about 2,000 armed vigilantes rounded up strike supporters on the morning of 12 July 1917. The vigilantes marched the men to the local baseball park, where they thinned the ranks of the captives to about 1,200 and then loaded them onto railroad cattle cars and deported them to Columbus, New Mexico, 173 miles east of Bisbee. Fewer than half of those deported were striking Wobblies. The roundup was merely an excuse to expel members of the MMSW local and anyone sympathizing with the strikers or opposing the mining companies. The vigilantes then established a kangaroo court to rid the Bisbee-Warren district of the remaining "radicals" while refusing those already deported the right to return to their homes. These controls remained in place until November and aided in the suppression of the strike.[83]

By the end of 1917, a combination of forces had practically eliminated organized labor in the Arizona copper districts. The internal conflicts within organized labor, especially between the MMSW and the IWW, made effective action difficult, and racial and ethnic divisions remained a major obstacle. Federal policies, including the continued use of the military to maintain "law and order" in the mining camps throughout the war, further weakened labor. Except for a few company unions that appeared in the early 1920s, labor unions disappeared in the Arizona copper industry until the late 1930s.[84]

Arizona was not the only western state to suffer from violent labor-management relations in the early twentieth century. Nevada copper workers established local unions affiliated with the WFM in 1906 and 1908, with the first serious strike, in July 1909, against the Cumberland-Ely Company. This well-organized effort continued for two years, with the union winning pay increases but not recognition by the company. The Western Federation locals struck the Nevada Consolidated Mining Company's operations in Nevada on 2 October 1912 and began a strike against all the operators on 14 October. Company gunmen killed two Greek workers on 17 October, prompting Governor Tasker L. Oddie to declare martial law and attempt to arrange a settlement. The strike ended on 28 October, when Nevada Consolidated granted a pay increase it had offered a month earlier but refused to recognize the union. The Nevada copper district remained calm until 29 July 1919, when the MMSW struck the Nevada Consolidated mines, demanding a wage increase of a dollar a day. The union settled for seventy-five cents a day on 29 July following an uneventful strike.[85]

Labor-management relations were much worse at the Utah Copper Company open-pit mine at Bingham Canyon, Utah. The mine had an ethnically diverse group of workers—including many Greeks, Italians, Austrians, and Japanese—which made organizing difficult. The first significant labor unrest at Bingham began on 30 September 1909, when three hundred Greek workers struck Utah Copper for a nine-hour workday and a pay increase of fifty cents a day. The weak WFM local aided their efforts, which resulted in a pay raise of twenty-five cents a day.[86]

A strike at the American Smelting and Refining Company lead smelter at Murray, Utah, beginning on 1 May 1912, set the tone for a major strike at Bingham Canyon later in the year. The WFM called the walkout, which lasted only six weeks before Greek strikebreakers brought it to an end. The Greeks came on the orders of Leonidas G. Skliris, a leading western labor agent known as the "Czar of the Greeks." Greeks worked and lived in Bingham under the *padrone* system originally developed by Italian labor agents. Besides extorting bribes from his countrymen before finding them jobs, Skliris also threatened them with discharge if they did not trade at his Pan Hellenic Grocery store. When the WFM began to organize the Utah Copper workers in the summer of 1912, Greeks formed the core of its support.[87]

The Western Federation began organizing at Bingham Canyon in the summer of 1912, had 250 members in July, 900 by the end of August, and about 2,500 in October. Despite pleas by federation president Charles W. Moyer for the local to negotiate with the Utah Copper officials, the men voted unanimously on 17 September to strike for a wage increase and union recognition. A potentially dangerous confrontation developed early in the strike when about 800 strikers, mostly Greeks, entrenched themselves in the mountainsides around the open pit, armed with rifles. They fired down into the mine workings, making any effort to operate equipment dangerous. On 20 September a Greek Orthodox priest persuaded the men to come down from their perches. The men then offered to return to work at the old pay scale if Utah Copper would sever all relations with Skliris.[88]

Although the "Czar of the Greeks" resigned from his post as labor recruiter for Utah Copper on 24 September, the union continued the strike. By the middle of October, the mine owners had imported about 5,000 strikebreakers, mostly from Mexico, and had resumed work on a limited basis. At the end of the month, Jackling announced a "voluntary" pay increase of twenty-five cents a day, half the increase demanded by the WFM. By mid-November the Bingham mines were gradually returning to normal operation and the strike was effectively dead. The only permanent improvement that the workers enjoyed because of the strike was the end of the power of Leonidas Skliris and the exposure and weakening of the padrone system in the Utah mines. Unions were nonexistent in Bingham Canyon until an abortive strike in 1931, followed in October 1936 by a successful one.[89]

Organized labor had little long-term success in the western copper districts, except at Butte. The absentee mine owners were more intransigent in dealing with organized labor than were the resident owners in Montana, and more willing to use violence to retain control. The growth of open-pit mining made organizing even more difficult in the West because skilled workers were less important than in underground mining.

CHAPTER 7

From the 1920s to the Vietnam War

Copper no longer is undisputed king among nonferrous metals from the standpoint of tons of metal produced from ore each year. . . . Officials of some of the companies that produce large quantities of copper express grave concern over the growing tendency to substitute other materials for uses in which copper heretofore has been predominant.—A. B. Parsons, *The Porphyry Coppers in 1956* (1957)

The decade of the 1920s was, in retrospect, a permanent watershed for the United States copper industry. The collapse of world copper prices in 1920–21, following the artificially high prices of the war years, marked the end of an era of easy growth and profits for American copper companies. As production costs increased at most American mines and new, low-cost producers emerged around the globe, the dominant position of the American copper companies in world markets steadily eroded. The U.S. share of world copper production, which stood at between 55 and 60 percent between 1900 and 1920, gradually fell to 48 percent in 1929 but rapidly after that. By 1960 the United States was producing only 24 percent of world copper, and by 1986 a mere 14 percent.[1]

Any division of the long time span from the early 1920s to the present would be inherently artificial. Historically, the U.S. copper industry weathered sharp fluctuations in demand resulting from the business cycle, the demands of war, the emergence of new producing regions, and changing patterns of copper consumption at home and abroad. The incomplete economic recovery of the 1920s, followed by the Depression, World War II, and the Korean War created particularly unstable conditions over an extended period. Following unsuccessful efforts to diversify between 1950 and 1957, most of the leading American companies refocused their resources on copper. Their best efforts resulted only in a production plateau of 1.5 million tons per year from 1957 through 1963. Global production and consumption grew rapidly by more than a third during the same period, so the U.S. copper industry continued its decline relative to the rest of the world.

From the mid-1960s on, the copper industry underwent substantial, permanent changes. The principal firms diversified and became targets for takeovers by various oil companies. The companies with large foreign holdings, particularly Anaconda and Kennecott, lost those properties via nationalization or expropriation. The 1970s and 1980s brought declining production and profits, the closing of dozens of mines, and a sharp reduction in employment. Labor-management relations deteriorated and in many respects returned to the conditions prevailing before unionization.[2]

This chapter examines significant changes in the American copper industry between the early 1920s and the early 1960s. There were fewer independent copper companies at the end of the period than at the beginning, but the survivors were more vertically integrated and simultaneously more diversified than before. By the 1960s, open-pit mining had replaced underground operations throughout the industry. Given the dominance of multinational firms, the regional framework used to examine the copper industry before 1920 becomes largely irrelevant. Changes in the copper industry—including fluctuations in demand, supply, and prices—must be understood within national and international contexts.

Mergers, Integration, and Diversification

In the post–World War II period, six large firms dominated the U.S. copper industry: the Anaconda Company, the Phelps Dodge Corporation, the Kennecott Copper Corporation, ASARCO (the American

Smelting and Refining Company) Incorporated, the Newmont Mining Corporation, and AMAX, established by a merger of the American Metal Company and the Climax Molybdenum Company in 1957. By the early 1970s, the six firms accounted for 72 percent of American output, with the three largest—Kennecott (25 percent), Phelps Dodge (18 percent), and Anaconda (12 percent)—producing nearly half. Between 1920 and the early 1960s, all six moved simultaneously toward greater vertical integration of their copper ventures and diversification into new endeavors. Five of the six became significant multinational enterprises by the 1950s. Phelps Dodge was the lone exception.[3]

By the late 1920s, the Anaconda Company had already achieved considerable vertical integration. In 1914, Anaconda gained control of the International Smelting & Refining Company, which owned four large copper and lead smelters and refineries. Anaconda's most daring acquisition came in 1922 when it bought the American Brass Company, the dominant fabricator of brass products in the United States. Anaconda's second major move into fabrication was the creation of the Anaconda Wire & Cable Company in 1929–30, combining its own wire manufacturing capacity with two departments of American Brass and six independent wire and rubber insulation manufacturers. Anaconda Wire & Cable began as the second largest manufacturer of wire products in the United States.[4]

Anaconda also became a more diversified metals company. It had been producing zinc at a large plant in Great Falls, Montana, since 1916 and was a substantial producer of magnesium during World War II. Acting in effect as an arm of the U.S. Atomic Energy Commission, Anaconda also developed a uranium mine and processing plant near Grants, New Mexico, between 1951 and 1953. The company's greatest single blunder of this period was its decision in 1951 to produce aluminum in competition with the three well-established firms already in the industry—Alcoa, Reynolds, and Kaiser. In 1955, Anaconda opened an aluminum processing plant at Columbia Falls, Montana, near a new federal hydroelectric plant, but it was never able to produce aluminum at a profit.[5]

The fundamental problem that Anaconda faced in the years following World War II was the depletion of ore reserves at Butte. In 1948 the company launched its Greater Butte Project to mine the large low-grade ore deposits underlying Butte through block-caving methods. Anaconda also opened the Berkeley Pit on the eastern edge of its Butte properties in 1955, but neither of these ventures was profitable.[6]

Anaconda increasingly relied on its overseas properties. By 1956, Anaconda's domestic mines, mainly at Butte, were yielding 128,000 tons of metallic copper per year. The Chilean mines produced 309,000 tons, and the Cananea mines in Mexico an additional 34,000 tons, or 73 percent of total production. The loss of its Chilean properties through expropriation in the early 1970s was a financial disaster for Anaconda and ultimately led to its demise as a copper producer.[7]

Phelps Dodge adopted strategies for survival that contrasted sharply with those followed by Anaconda but that ultimately proved more successful. The firm's involvement in foreign copper mines remained minor. Phelps Dodge instead moved toward greater vertical integration in 1930, when it gained control of the Nichols Copper Company and built a new refinery at El Paso, Texas. In the fall of 1930, Phelps Dodge also bought the National Electric Products Corporation, a large U.S. manufacturer of copper wire and cable.[8]

Starting in the 1920s, Phelps Dodge gained complete control over both the Clifton-Morenci and Bisbee-Warren districts in Arizona. The need to control vast amounts of ground to develop open-pit mines in order to exploit low-grade deposits prompted expansion. The major acquisitions were the Arizona Copper Company (1921) at Clifton and the Calumet & Arizona Mining Company (1931), with mines at Bisbee and Ajo, Arizona. Finally, the company bought the United Verde Copper Company from the heirs of William A. Clark in 1935 for $10 million in cash.[9]

Phelps Dodge worked two open-pit mines at Bisbee, but the deposits were too small for long-term success. The first, which operated from 1923 to 1932, was a low-grade deposit that outcropped to form Sacramento Hill southeast of Bisbee. The second, the Lavender Pit southeast of Sacramento Hill, was the result of the U.S. government's efforts to increase copper production in the early 1950s. The federal government gave Phelps Dodge a five-year write-off of its investment in this venture and guaranteed a base price of 22 cents per pound for 112,000 tons of metallic copper. The Lavender Pit went into operation in 1954 and ran for twenty years but earned only modest profits for most of its life.[10]

Exploratory drilling at the Clay Orebody at Morenci between 1928 and 1930 revealed roughly 200 to 300 million tons of ore containing 1 percent copper, but economic conditions in the early 1930s ruled out any development. The company began developing an open-pit mine at Morenci in June 1937 and began work on a reduction works in 1939 with a daily capacity of 27,000 tons of ore. With the growing threat of war, the U.S. government asked Phelps Dodge to increase the plant capacity to 42,000 tons and invested $26 million through the Defense Plant Corporation to build additional capacity. Phelps Dodge spent $42 million of its own funds on the Morenci project, making a total investment of $68 million.

When Morenci came on line in early 1942, Phelps Dodge's copper production jumped by one-third. More important, further explorations had revealed that the Clay Orebody contained 450 million tons of 1 percent ore and an equal amount of even lower grade material. At the end of the war, Phelps Dodge bought the parts of the Morenci complex built by the Defense Plant Corporation. By 1948, Morenci had a daily capacity of 50,000 tons of ore and was the most efficient copper mining/smelting plant in the world. By then, Morenci had become the centerpiece of the Phelps Dodge copper empire, and in many respects, it still is to this day. By developing Morenci and expanding its other Arizona mines, Phelps Dodge could substantially increase output over the long run. The company had produced about 100,000 tons per year from 1925 to 1929, reached a wartime peak of 239,000 tons in 1943, but then averaged 242,000 tons of copper per annum from 1947 to 1950, a period of little military demand.[11]

Another large producer was the Kennecott Copper Corporation, which Daniel Guggenheim established in April 1915 to unite the Guggenheim family mining properties. The new corporation was initially an investment company, not an operating company. Its assets included the Kennecott Mines Company, which owned several Alaska properties; 98.5 percent of the shares of the Braden Copper Company, which operated the Braden (El Teniente) mine in Chile; and one-quarter of the shares of the Utah Copper Company, which owned the Bingham Canyon mine. For unknown reasons, the Guggenheims did not include the Chile Copper Company, which controlled the Chuquicamata mine, in the Kennecott Copper Corporation.[12]

Kennecott made only minor moves toward increased vertical integration until Charles R. Cox, previously with U.S. Steel, became its chief executive officer in 1950. Historically, Kennecott had relied on ASARCO to smelt its ores, in large part because the Guggenheim family controlled both firms. Kennecott and ASARCO even shared the same office building until 1950. In 1958, under Cox's leadership, Kennecott bought ASARCO's Garfield plant near Bingham Canyon and built a new smelter in Hayden, Arizona, to handle ore from the Ray mine. The following year, Kennecott opened a new refinery in Baltimore to refine the product from Braden, previously handled by ASARCO.

Kennecott also diversified its interests, mostly through joint ventures with other companies. In 1945 it engaged in petroleum exploration on a joint venture basis with Continental Oil but with little success. The copper giant also diversified into aluminum in 1953, buying a 13 percent interest in Kaiser Aluminum. In 1948, in partnership with New Jersey Zinc, Kennecott established the Quebec Iron & Titanium Corporation, but this venture barely broke even during its first decade of operation and then made only modest profits. By the late 1950's, Kennecott has lost its appetite for diversification.[13]

The American Smelting and Refining Company (ASARCO), did not mine copper in the United States in a substantial way until the 1950s. ASARCO was the third largest copper processor in the United States in 1946 and led the world in treating gold, silver, lead, and zinc. Although it was almost exclusively a refining and smelting firm for much of its history, ASARCO was a leading player in the U.S. copper industry because of its work for other firms and its ties to the Guggenheims. Following the tangled tendrils of finance and management connecting ASARCO, Kennecott, and various other Guggenheim enterprises is no simple task.[14]

Meyer Guggenheim entered the field of mining by accident. In 1880 he invested in a Leadville, Colorado, silver mine, which was so successful that it made Guggenheim a millionaire. He then built his own reduction works at Pueblo, Colorado, with four of his seven sons as equal partners. In 1892, two of them, Daniel and Murry, bought a copper mine in Mexico and built two smelters there to process copper and other ores. The Guggenheims subsequently established the Guggenheim Exploration Company (named Guggenex after its cable designation) in 1899 to locate and develop mines around the world. Guggenex eventually developed copper mines in Utah, Nevada, Chile, and the Belgian Congo.

An alliance of smelter operators and financiers launched the American Smelting and Refining Company (ASARCO) in April 1899, intending to create a combine to control the smelting of lead and silver ores. ASARCO initially merged eighteen firms that controlled sixteen smelters, eighteen refineries, and several small mines but did not include the Guggenheims. After ASARCO got off to a shaky start, its owners negotiated an agreement that gave the Guggenheims effective control of ASARCO. Five Guggenheim brothers together held one-third of the fifteen directors' slots. The deal did not include Guggenex.[15]

The Guggenheims and their close associates controlled ASARCO from 1901 through 1957. Over time, the company engaged in forward integration into finished products, backward integration into mining, and general diversification. ASARCO built a brass and copper rolling mill in 1915 and a rod and wire plant in 1923, both in Baltimore, and then took over the General Cable Company in 1927. The following year, ASARCO bought the Michigan Copper & Brass Company and gained partial control over additional brass manufacturers, including the Taunton-New Bedford Copper Company, headed by E.H.R. Revere, the great-grandson of Paul Revere. ASARCO combined these companies and named the new venture the Revere Copper and Brass Company, evoking the memory of the famous patriot.

ASARCO also integrated backward into metal mining but did not acquire copper mines until the 1950s. By 1946, in the United States and Mexico combined, ASARCO owned a total of six copper smelters, eight lead smelters, two zinc smelters, and eleven refineries treating the three metals. ASARCO developed the Silver Bell mine near Tucson, Arizona, which it opened in 1953, and the Mission mine, also near Tucson, which went into production in the mid-1960s. The Silver Bell (with an annual capacity of 18,000 tons) and Mission (with a capacity of 45,000 tons) were small and mid-sized mines respectively.[16]

The remaining large firms, Newmont and AMAX, emerged as major producers in the 1950s. William Boyce Thompson, who developed the Shannon mine at Clifton, Arizona, established the Newmont

Mining Corporation in 1921 as a holding company for his various mining stocks. The name was a combination of the two places where Thompson spent much of his life, New York and Montana. Newmont owned stock in various metal mining companies, but over time, copper properties dominated the portfolio. Through the 1920s, Newmont made money simply by speculating in copper and other mining company shares but did not involve itself in managing any of the properties. Two African ventures were exceptions. In 1928, Newmont bought half the stock in the O'Okiep mine in South Africa. This copper mine shut down for most of the 1930s but earned healthy profits for Newmont over the long run. The company also bought a 30 percent interest in the Tsumeb mine in South-West Africa in 1947. This property was originally a high-grade surface copper mine developed by the Germans in the 1890s, but Newmont redeveloped it as a shaft mine. Newmont was also a partner in the Southern Peru Copper Corporation in the mid-1950s, with ASARCO as the principal investor.

Newmont became seriously involved in copper mining in the United States in the 1950s and then only reluctantly. In 1928 the firm bought William Boyce Thompson's shares in the Magma Copper Company, which operated a mine at Superior, Arizona. In 1945, Magma acquired a large low-grade copper deposit at San Manuel, Arizona, but was unable to raise the capital needed to develop the property. Magma made little progress at San Manuel until 1952, when the Reconstruction Finance Corporation (RFC) agreed to lend Magma $94 million to develop the property. The mine went into production in 1956 using block-caving methods but did not earn profits until 1960. The RFC projected an annual output of 70,000 tons of metallic copper, and San Manuel was producing roughly that amount by the early 1970s, making it the second largest copper mine in the United States, after Bingham Canyon. Once San Manuel began to earn profits, Newmont bought additional shares in Magma, and by 1969 the firm had obtained all of Magma's outstanding stock.[17]

The last of the major firms considered here is AMAX, established in 1957 by the merger of the American Metal Company and the Climax Molybdenum Company. The new firm was initially called the American Metal Climax Company, then AMAX in 1974. The American Metal Company began in 1887 as a metal trading firm, but by the late 1910s it controlled a half dozen lead, zinc, and copper refineries. In 1918, American Metal and its executives launched Climax Molybdenum, which was a leading producer of molybdenum until the early 1980s. The firm jumped feetfirst into African copper mines in 1927, initially by taking a minority ownership in the O'Okiep mine in South Africa developed by Newmont. American Metal then made large investments in Northern Rhodesia's Copper Belt, beginning in 1927 with the Roan Antelope mine, which proved to be an enormously rich property. Following World War II, American Metal had a 30 percent interest in the Tsumeb mine in South-West Africa, in a joint venture with Newmont. By 1965, AMAX's assets equalled those of ASARCO and Phelps Dodge, although its production of copper still did not.[18]

The two largest Michigan companies—the Calumet and Hecla Mining Company and the Copper Range Consolidated Mining Company—had become only mid-sized operators by the 1920s, but they imitated the behavior of their larger competitors. Between 1909 and 1911, Calumet and Hecla bought a controlling interest in most of its neighboring copper companies, intending to create a single consolidated operation, but stockholders' lawsuits delayed the planned merger until 1923. The company did not actively pursue vertical integration until 1942, when it bought the Wolverine Tube Company of Detroit, a large manufacturer of seamless copper pipes. After a group of Chicago investors took control of the company from the Agassiz and Shaw families of Boston in 1953, Calumet and Hecla diversified into timber, lumber, and plastics. Copper Range, which had begun as a holding company, gained direct control of its constituent firms in the 1920s, and in 1931 it bought C. G. Hussey & Company, which operated copper and brass mills in Pittsburgh. In the early 1950s, Copper Range used an RFC loan to open the White Pine Mine to exploit low-grade sulphide ores at the southern edge of Michigan's copper range.[19]

Foreign Adventures and Foreign Investments

The large, integrated copper companies that emerged in the early twentieth century made substantial investments in foreign mines.

The Guggenheims and Phelps Dodge developed mines in Mexico, while Kennecott and Anaconda invested heavily in Chile. Foreign expansion allowed these companies to expand profits by developing production in low-wage undeveloped countries, particularly Mexico, Chile, Peru, and Rhodesia. The development of foreign mining properties seemingly guaranteed a steady supply of copper at a time when U.S. copper reserves appeared fixed in size and subject to exhaustion within a few decades.

Mexico was the first target for U.S. investments in copper. Porfirio Díaz, Mexico's president from 1876 to 1911, encouraged foreign capitalists to invest in Mexico as a way to industrialize his poverty-stricken country. Daniel Guggenheim approached Díaz and won concessions to build three smelters in Mexico. In 1892, the Guggenheims opened a lead smelter at Monterrey and a year later a copper/lead smelter at Aguascalientes in the central part of the country. In 1902–3, the Guggenheims invested another $18 million in six mines in northern Mexico.[20]

Phelps, Dodge and Company developed its first Mexican mine, the Moctezuma, near the town of Nacozari, fifty miles south of the Arizona border. Louis D. Ricketts visited the Moctezuma in 1895 for Phelps Dodge and urged his employer to buy the property. Phelps Dodge followed his recommendation in 1897 and named Ricketts the general manager. By 1909 the Moctezuma mine was producing 13,000 tons of copper per year, about one-third of the output of the Copper Queen and one-fifth of the total for Mexico. Moctezuma remained a productive mine through World War II, when the ore reserves began to run out. Mining stopped for good in May 1949, but Phelps Dodge maintained a leaching plant at Nacozari through the end of 1960, when it too shut down.[21]

A second significant copper venture in northern Mexico began after "Colonel" William C. Greene (1853–1911), a Tombstone, Arizona, ranch owner, visited the abandoned Cananea copper mine in 1898 and launched the Cobre Grande Copper Company to develop the property. After fleecing the original investors, Greene organized the Cananea Consolidated Copper Company in September 1899. He persuaded the Amalgamated Copper Company, among others, to invest in his company. In 1906 the Amalgamated interests forced

Greene out of the company, which they reorganized as the Greene-Cananea Consolidated Copper Company. With Louis D. Ricketts in charge, production rose to 3,000 tons of copper per month in late 1911, and the mine paid its first dividend in March 1912.[22]

The Anaconda Copper Mining Company, the descendant of the Amalgamated, kept control of the Greene-Cananea Copper Company. In the mid-1920s, Anaconda found a significant new deposit, the Colorado orebody, which yielded considerable copper until its exhaustion in 1944. The subsequent exploitation of two large porphyry deposits by open-pit methods extended the life of Anaconda's Cananea operations. In 1956, two open pit mines and one underground mine yielded more than 33,000 tons of metallic copper, roughly half of the total Mexican production.[23]

Most American copper ventures in Mexico were less successful than the mines at Nacozari and Cananea. The Mexican Revolution (1911–17) brought disruptions and losses for most U.S. mining operations in Mexico. The American-owned Minas de Corralitos in the state of Chihuahua, for example, shut down in November 1911 following looting by Mexican federal forces and bandits. Mexican federal forces returned there in December 1917 and systematically demolished and pillaged the Minas de Corralitos properties, including the mine, concentrator, smelter, and townsite. The mines remained closed, and the American owners never received compensation for their losses.[24]

U.S. investment in copper mines and smelters in Mexico did, however, bring the rapid development of a modern copper industry starting in the 1890s, a development that would not have happened without American capital and technical expertise. The production of copper in Mexico roughly quadrupled between 1891 and 1900, from 6,200 tons to nearly 25,000 tons, and then increased to 72,000 tons in 1905, when Mexican output was one-eighth that of the United States. Following the ouster of Díaz in 1911 and a series of revolutions and political upheavals extending to 1920, the U.S. companies lost much of their privileged status and easy profits but continued to operate mines and smelters.[25]

American copper companies also made large investments in Chilean mines in the early twentieth century and earned impressive returns. The Guggenheims and Anaconda were the leading actors. The first significant new copper property developed in Chile in the early twentieth century was the El Teniente mine, located in central Chile roughly 9,000 feet above sea level. In October 1903, Marco Chiapponi, an Italian-Chilean mining engineer, directed the mining engineer William Braden toward these deposits. Chiapponi urged Braden to find U.S. investors to develop the property, which included at least 250,000 tons of ore with a copper content of 4.5 percent. After visiting the mine location, Braden quickly raised $30,000 from several top executives in ASARCO and began buying the property. Braden and his partners then organized the Braden Copper Company.

Their initial stock offering raised $625,000, which financed the purchase of the property and the initial development. By June 1906 the mine was working, with a 250-tons-per-day concentrating plant in operation, but Braden needed more capital to build a smelter and a railroad. The Guggenheim Exploration Company bought $500,000 of Braden Copper Company bonds in 1907 and received a mortgage on the property in return. In June 1909 the Guggenheims supplied $4 million in working capital through a bond issue and thus gained control of the Braden mine. The Kennecott Copper Corporation subsequently bought the stock of the Braden Copper Company in 1915, paying $42 million in Kennecott shares. Because Kennecott also assumed $15 million in outstanding debts, it paid $57 million for Braden.[26]

Braden was the first porphyry mine in Chile and was an enormous success. The production of metallic copper, a mere 1,500 tons in 1908, reached 8,000 tons in 1913 and then climbed dramatically to 39,000 tons in 1918. Output doubled again by 1924 to 78,000 tons and then peaked at 109,000 tons in 1928, making Braden the third largest porphyry mine in the world. By 1929, Kennecott had extracted 45 million tons of ore from Braden, and it estimated the remaining reserves at 350 million tons of 2.18 percent ore. Braden remained in the hands of Kennecott until the Chilean government nationalized the property in 1967 and then expropriated it four years later.[27]

The second large porphyry operation in Chile was the Chuquicamata mine, located in a desert region in northern Chile near the border with Bolivia. One of Daniel Guggenheim's engineers had visited this area in 1900 and had urged him to buy this property, but Guggenheim had little interest in a deposit of 2 percent copper ore at a time when Jackling had not yet proved the value of such low-grade ores. Ten years later, Albert C. Burrage, a Boston financier, acquired options on the properties in the area and approached the Guggenheims for financial support, convinced that this copper deposit could be mined profitably.

After an independent evaluation, Daniel Guggenheim concurred and established the Chile Exploration Company in January 1912 to develop the deposits. The new company proceeded with systematic exploration by means of churn-drilling. In January 1913 the engineers estimated that the deposit contained at least 75 million tons of ore averaging 2.8 percent copper. By late 1916 this estimate had increased to 700 million tons averaging 2.12 percent copper. The Guggenheims incorporated the Chile Copper Company of Maine in December 1912 with capital stock of $20 million, but replaced it with the Chile Copper Company of Delaware, incorporated in mid-April 1913 with a capital stock of $95 million, soon raised to $110 million. Burrage received nearly one-quarter of the stock. The Chile Exploration Company, wholly owned by the Chile Copper Company, became the operating entity that managed the Chuquicamata.[28]

Developing these deposits in such a remote part of Chile proved expensive and time-consuming. This was an open-pit operation from its conception, using steam shovels for excavation and steam railroads for transportation. Chile Exploration bought the steam shovels used by the U.S. Army engineers to dig the Panama Canal. It built a 53,000-horsepower oil-powered electric generating plant costing $3.5 million at the port of Tocopilla, some ninety miles from the mine site. In total, Chile Exploration spent over $12 million to bring the mine into operation. Chuquicamata began production in April 1915 with a 10,000-tons-per-day crushing plant, enlarged by the late 1920s to a capacity of 55,000 tons.

The Guggenheims needed additional capital to develop their interests in Chile's nitrate industry, so they sold a controlling interest (53 percent) in Chuquicamata to the Anaconda Copper Mining Company in 1923 for $70 million. Opposition to the sale led two Guggenheims to resign from the firm. In early 1929, when world copper prices were peaking, Anaconda bought most of the remaining shares for roughly $220 million in Anaconda stock. The production record of Chuquicamata was impressive. Metallic copper output was 21,000 tons in 1916, the first full year of operation, but jumped to 107,000 tons in 1924 and peaked at 149,800 tons in 1929.[29]

The last large porphyry deposit developed in Chile before World War II was the Andes mine at Potrerillos, roughly 300 miles south of Chuquicamata. William Braden took options on the property in May 1913 and began exploratory churn-drilling. After Braden convinced the Anaconda Copper Mining Company of the potential of the Andes mine, Anaconda organized the Andes Copper Mining Company in January 1916 with a nominal capital stock of $50 million. Braden transferred his options on roughly 3,500 acres, along with water and railroad concessions, to the new company in return for a block of shares. Additional exploration revealed a substantial deposit of more than 100 million tons of ore averaging 1.50 percent copper, considerably leaner than the ores at Braden and Chuquicamata.

Anaconda began to develop the mine in 1917, but the sharp drop in copper prices following World War I prevented further progress for six years. Once world copper prices improved, however, Anaconda went ahead with full-scale development in early 1925. Anaconda invested more than $47.5 million in bringing the mine into full production, with the bulk of the spending concentrated in the period 1925 to 1928. The Andes mine produced 27,200 tons of metallic copper in 1927, the first year of operation, and a remarkable 81,300 tons in 1929, only the third year of production. It ranked as the fourth largest porphyry mine in the world in 1929 and approached the output of Braden.[30]

The El Teniente, Chuquicamata, and Potrerillos (or Andes) mines transformed and revitalized Chile's mining industry and accounted for 90 percent of Chile's copper production in 1929. The newest of the three, the Andes mine, was the first to exhaust its ore reserves. Production exceeded 100,000 tons in 1941, but by the late 1940s the Andes was in decline. When it shut down in 1958, the copper content of the ore had dropped to 0.6 percent.

Anaconda purchased claims in the Indio Muerto (Dead Indian) Mountain district fifteen miles northwest of Potrerillos and in 1952 began a diamond drilling program that discovered an orebody of more than 200 million tons averaging 1.6 percent copper. There, Anaconda developed a new mine, named El Salvador, which opened in 1958 as a replacement for the Andes mine. Anaconda spent over $100 million to develop the new property as an underground block-caving operation and built the new company town of El Salvador but continued to operate its smelter at Potrerillos.[31]

U.S. investors also developed and controlled the chief copper mines of Peru through most of the twentieth century. In 1901, James B. Haggin and the estate of George Hearst formed a syndicate to buy mines in the Cerro de Pasco region in the central highlands of Peru. The resulting Cerro de Pasco Corporation quickly gained control of more than a dozen mines producing gold, silver, and copper. Cerro de Pasco dominated Peruvian mining between 1922 and 1928, when it produced more than 80 percent of the total output of the nation's metal mining industry. However, Cerro de Pasco's production of copper remained under 50,000 tons per annum from the 1920s through the late 1960s, making it at best a mid-sized copper company on the world scene.[32]

ASARCO was also involved in Peruvian copper mining. It established the Northern Peru Smelting & Refining Company in 1921 as a wholly owned subsidiary to develop mines in northern Peru. The firm had only mixed results with copper during the 1920s.

Following World War II, ASARCO purchased a large mine property at Toquepala in southern Peru. Exploratory drilling that started in 1948 revealed a deposit of 1.25 percent copper sulfide ore best exploited by open-pit methods. The American smelting giant lacked sufficient capital and expertise in open-pit mining to develop the property on its own, so in March 1952 it approached the U.S. government's Defense Materials Procurement Agency, requesting a loan of $120 million to develop the mine. The smelting firm would provide $46 million of its own capital for the project. The American government supported the proposal as a means of bolstering the supply of this critical metal for the U.S. market and as a way of promoting economic development in Latin America. In April 1953 the U.S. government was prepared to grant a $60 million loan through the Defense Production Administration and a second $60 million loan through the Export-Import Bank. Revised estimates of development costs came to $177 million, however, forcing a new set of negotiations.

In early 1955, ASARCO created a joint venture, the Southern Peru Copper Corporation (SPCC), with Cerro de Pasco, Phelps Dodge, and the Newmont Mining Corporation as minority owners. ASARCO and its partners agreed to spend at least $95 million of their own capital and received a $100 million loan from the Export-Import Bank. The SPCC began development work in 1956 and spent nearly $250 million before the mine opened in 1959. Toquepala produced more than 145,000 tons of copper in 1960, the first full year of operation, and 160,000 tons in 1961. Output then averaged about 130,000 tons per year for the rest of the decade. With total costs of production per pound of copper roughly half the selling price, Toquepala generated large profits for ASARCO and its partners.[33]

Midsized American mining companies invested in metal mining in Africa. The Newmont Mining Corporation and AMAX developed the O'Okiep mine in South Africa in 1927 and the Tsumeb mine in Southwest Africa in 1947. AMAX and the Rhodesian Selection Trust (RST) opened the Roan Antelope and Mufulira mines in Northern Rhodesia in the late 1920s and 1930s; and the Anglo-American Corporation of South Africa invested in copper mines in Northern Rhodesia in the mid-1920s. The largest American copper companies of the 1920s—Anaconda, Kennecott, and Phelps Dodge—had no involvement in developing the Rhodesian copper deposits, because their geologists and engineers did not believe that the deposits would ever be commercially viable.[34]

The U.S. Copper Industry in An Era of Economic Challenge

Over four decades starting in the early 1920s, the U.S. copper industry had to weather one of the most unstable economic environments of its entire history. The industry's recovery from the 1920–21 worldwide collapse of copper prices and production was unsteady and gradual, with full prosperity not regained until 1929, on the eve of the Great Depression. The industry then slumped badly until the late 1930s, when wartime demand brought some relief. In late 1929 the

British government ordered the closure of the London Metal Exchange, which had served as the primary international market for copper futures since 1877. The rapid growth in demand for copper during World War II and the Korean War further disrupted copper markets, brought the imposition of government economic controls, and ended with years of painful readjustment to a "peacetime" economy. The British government finally allowed the London Metal Exchange to reopen in 1953 after a fourteen-year hiatus.[35]

A profound shift in the industry's international position in part reflected the sluggish growth in U.S. output. American copper production rose from 837,000 tons in 1925 to 1,080,000 tons in 1960, an increase of only 29 percent. In contrast, world copper production more than tripled between 1925 and 1960. The U.S. domestic copper industry produced roughly 55 to 60 percent of world output between 1900 and 1920 and still accounted for 48 percent in 1929, the only prosperous year of the entire decade. Its share of world output then plummeted to 24 percent in 1960. The decline was less sharp if we include the output of the Chilean mines entirely owned by American companies.[36]

The relative stagnation of the U.S. domestic copper industry resulted from several interconnected economic trends—the sluggish growth of domestic demand for copper; the loss of export markets to cheap foreign copper; the industry's failure to develop significant new mines within the United States; and the decisions by the large multinational mining companies to direct their investments into their overseas ventures. The failure of American companies to open new mines in the United States or to develop new cost-cutting technologies resulted in higher production costs, the loss of markets, and reduced profits. These conditions in turn discouraged additional investment within the United States.

World copper production had grown at a robust rate of 5.8 percent per year over the period 1890–1918, reflecting the rapid industrialization of North America and western Europe. The long-term growth in world copper production slowed considerably after World War I to roughly 3.0 percent per annum over the period 1918–1950, a reflection of the severe depressions that the world copper industry experienced in 1920–21 and in the 1930s. Still, the demand for copper continued to expand steadily after World War I.

TABLE 7.1
Consumption of Copper by Principal Users as Proportion of Total Consumption, 1920–1950

	1920	1929	1950
Electrical Manufactures	26.1%	22.5%	20.0%
Telephones and Telegraphs	7.7	14.4	6.0
Light and Power Lines	14.6	11.0	7.0
Railways and Ships	9.8	1.4	4.0
Automobiles	7.6	11.9	10.0
Buildings	4.1	5.1	10.0
Radio and Television	0.0	1.3	4.0

Sources: Mineral Industry, 1921, 148; *Mineral Industry, 1931,* 148–149; and President's Materials Policy Commission, *Resources for Freedom,* vol. 2: *The Outlook for Key Commodities* (Washington, D.C.: USGPO, 1952), 33.

The 1890s marked the beginning of rapid electrification in the advanced countries, particularly in the United States. Copper became the raw material of choice for electric transmission lines and for wiring used in motors, generators, switching equipment, and telegraph and telephone systems. During the year ending 30 June 1895, the Western Union Telegraph Company replaced 10,000 miles of iron wire with copper. Copper used for electrical applications still accounted for 45 to 50 percent of copper consumption in the United States in the 1920s. Per capita consumption of copper also increased from 13.5 pounds in 1924 to a peak of 20.9 pounds in 1929, an impressive jump in only five years. After the 1920s, when the United States had completed the bulk of its electrification, domestic demand stagnated permanently.[37] The changing patterns of copper consumption can be seen in table 7.1.

Facing a slump in demand following the end of World War I, forty-two of the largest copper producers and brass manufacturers, led by Anaconda and Phelps Dodge, established the Copper & Brass Research Association in 1921 to encourage the consumption of copper and its alloys. The new trade association hired the public relations firm of Ivy L. Lee and Associates to develop a campaign to persuade consumers of the long-term cost advantages of copper over alternate materials. The association hired researchers to discover new uses for copper and, distributed a free copy of its monthly bulletin (published from 1921 to 1946) promoting the use of copper by the construction industry.[38]

Copper consumption in the United States temporarily increased during World War II, but in 1946, the first full year of peace, consumption stood at only 19.8 pounds per capita, below the 1929 level. In the postwar years, the copper industry faced increased competition from other materials, particularly aluminum. In the mid-1950s, local electrical utilities and American Telephone and Telegraph began substituting aluminum for copper in their lines.[39]

The increased supply of scrap copper that developed in the 1920s also further dampened the demand for "new" copper from the mines. During the decade 1911–20, copper produced from "old scrap" amounted to between 10 and 20 percent of new copper produced from the mines and smelters. In the years 1921–30, "production" from scrap copper was 35 to 50 percent of mine output. Even with the revival of mining in 1937, copper produced from scrap typically amounted to half the volume of "new copper" from the mines. Mature industrialized economies like the United States generate large quantities of scrap, which depresses copper prices.[40]

Following World War I, the porphyry mines of Chile and new mines in central and southern Africa flooded world markets with cheap copper. Between 1925 and 1960, when U.S. copper output rose by only 29 percent, Chile increased its production by 120 percent. The development of mines in the Katanga Province of the Belgian Congo (Zaire) in the early 1920s and the emergence of significant mines in Northern Rhodesia (Zambia) and in South-west Africa created a new major player in the world copper industry. African copper output jumped from 120,000 tons in 1925 to 1,076,000 tons in 1960, a ninefold increase. By 1960, African copper production had come to equal that of the United States.[41]

Starting in the 1920s, U.S. copper companies were hard-pressed to sell their copper overseas. Throughout the decade of the 1910s, the United States had enjoyed an export surplus in copper of roughly 200,000 to 250,000 tons per year. This surplus was cut in half in the 1920s and briefly disappeared from 1929 to 1933. Beginning in 1940, the United States permanently became a net importer of copper. During the years 1942 to 1945, the import surplus averaged more than 560,000 tons per year, or one-quarter of consumption.[42]

Long-term trends in the world copper industry discouraged American firms from investing in their domestic properties. The real selling price of copper (adjusted for inflation) fell by half over the long period 1890 to 1945 and then was stable from 1945 to the 1980s. The quality of U.S. ore deposits has steadily declined since World War I, driving up production costs and making further investment unattractive. The U.S. corporations with significant foreign holdings—particularly Anaconda, Kennecott, ASARCO, and Climax—instead invested in their overseas properties.[43]

Except for the Phelps Dodge open-pit mine at Morenci, which opened in 1942, no significant new copper mines went into production in the United States between 1920 and the early 1950s. The industry shut down older, less efficient underground mines, including Kennecott's Alaska mines (1938); the United Verde Extension in Jerome, Arizona (1938); the Old Dominion mine at Globe, Arizona (1941); the Quincy mine in Hancock, Michigan (1945); and the United Verde in Jerome (1953). Copper production in the United States increasingly came from the porphyry mines of the West. Arizona, Utah, Nevada, and New Mexico accounted for 52 percent of national output in 1911–15, but in 1951–55 the four states produced 87 percent of the total.[44]

At the end of the Great War, the U.S. copper companies found themselves with large stocks of copper and falling prices. The Webb-Pomerene Act of 1918 granted export associations (cartels) based in the United States limited exemption from antitrust laws. As a result, the large U.S. copper companies established the Copper Export Association (CEA) in December 1918 to dispose of their copper stocks slowly in order to reduce downward pressure on prices. The CEA reduced U.S. inventories from 562,000 tons in 1921 to 289,000 tons in 1923, when the association dissolved.

An effort to create a global copper cartel came in October 1926 with the formation of Copper Exporters, Incorporated (CEI), which included the large African mines. Its members accounted for about 85 percent of world output in 1927–29. The CEI restricted output during these years and thereby contributed to the price increases of the same years but was unable to prevent the disastrous slide of world copper prices in the early 1930s. U.S. companies, which had sold copper for 18 cents a pound in 1929, saw prices fall below 6 cents in 1932. When the American government protected domestic producers with a tariff of 4 cents per pound in 1932, the foreign copper companies withdrew from the CEI, effectively killing it.[45]

The post–World War I "recovery" was at best an anemic one. American copper output, which had peaked at about 950,0000 tons per year in 1916–18, fell to 600,000 tons in 1919–20 and to a mere 233,000 tons in 1921. Output then rebounded to 482,000 tons in 1922, climbed to 740,000 tons in 1923, and slowly increased over the rest of the decade. In 1929 the U.S. copper industry barely matched the peak wartime production of 1916, while world copper output was one-third larger. Most of the large companies enjoyed only modest earnings during the 1920s. Phelps Dodge earned profits only in 1928 and 1929, while Anaconda's earnings over the decade were a meager 2 percent of net assets. ASARCO's rate of return on investment for the decade was roughly 11 percent, and Kennecott earned a healthier 14 percent return.[46]

Despite stagnant prices and mediocre profits, the copper industry enjoyed relative prosperity in the 1920s compared to its fate during the 1930s. Prices bottomed out at less than 6 cents a pound in 1932 and recovered to only 9.5 cents in 1936. The boom of 1937 lifted prices briefly to 13 cents, but from 1938 to 1941, copper sold for only 10 to 11 cents a pound. American copper production fell to 190,000 tons in 1933, roughly the same output level as 1895. Production levels rebounded quickly to a peak of 842,000 tons in 1937 before dropping again in the late 1930s. The U.S. copper industry did not surpass its 1929 production level until 1942, when output reached 1,080,000 tons.[47]

Following the breakup of Copper Exporters, Inc., in 1932, international efforts to stabilize copper prices proved futile until March 1935, when a new cartel emerged. It included only five voting members—the two American companies with large foreign operations, Anaconda and Kennecott; Roan Antelope and Rhokana, both in Rhodesia; and Katanga (in the Belgian Congo). The Bor Mine (in Yugoslavia) and Río Tinto (in Spain) were nonvoting members, and other large foreign producers, including those in Canada and Peru, were "friendly" to the cartel. The official members of the international cartel accounted for only about one-third of global copper output, but when the various other firms associated with the cartel are included, the cartel typically controlled between 50 and 55 percent of world production. The international cartel remained in operation from 1935 through 1956, and its members effectively colluded to restrict output and stabilize prices over those years.[48]

The U.S. copper companies emerged from World War II in a weakened condition, a result of government policies and their own actions during the war. Besides price controls, the government imposed a stiff excess profits tax based on earnings in the late 1930s. This was a burdensome tax because most of the large companies had earned little or no profit during the base years. During the war the largest companies rewarded their long-suffering stockholders by paying out between two-thirds and all of their after-tax profits as dividends. The net worth of all the large copper firms declined in real terms during the war, and they entered the postwar period with almost no capital reserves for new investment.

The end of government controls in late 1946 brought a return of unstable conditions but also a return to profitability. Price subsidies expired in September 1945, and all price controls ended on 10 November 1946. Copper prices, which had averaged 13 cents a pound in 1946, then jumped to 23 cents in 1947. A series of strikes in 1946 reduced output from the 1944 level by one-third, but production of new copper then rebounded to about 840,000 tons in both 1947 and 1948. Congress suspended the 4-cent duty on imported copper starting on 30 April 1947, and except for a nine-month period in 1950–51, foreign copper paid no duties until July 1958. Imports, which averaged roughly 400,000 tons per annum in 1946–47, increased to an annual average of 583,000 tons in 1948–51. With the end of government controls, the leading copper companies again earned reasonable profits. The principal firms, except Anaconda and ASARCO, invested most of their postwar profits in unsuccessful oil exploration ventures.[49]

The demand for copper in the immediate postwar years would have been even more robust had the federal government carried out its plan to create strategic stockpiles of copper. In 1946, Congress authorized copper stockpiles of 1.13 million tons, increased in 1950 to 1.9 million tons. Because of budget limitations, however, copper stockpiles reached only 0.54 million tons in 1951. The start of the Korean War in June 1950 stimulated the industry—production jumped from about 750,000 tons in 1949 to 908,000 tons in 1950, and average prices increased from 20 cents a pound in 1949 to 21 cents in 1950. When prices rose sharply in the last half of 1950, the federal government placed a ceiling price of 24.5 cents per pound on domestic copper, effective January 1951. The ceiling price remained in place for all of 1951 and 1952. "Free market" prices in 1953 averaged just under 29 cents a pound.[50]

Concerns about shortages of copper and other strategic materials during the Korean War led President Harry S. Truman to create the President's Materials Policy Commission in January 1951 to recommend long-term solutions. The Paley Commission (as it was known, after its chairman, William S. Paley of the Columbia Broadcasting System) issued a five-volume report entitled *Resources for Freedom* in June 1952. A considerable part of the second volume, *The Outlook for Key Commodities,* dealt with copper and was pessimistic about the nation's future supply.

The Paley Commission estimated that copper consumption in the United States would increase roughly 45 percent between 1950 and 1974, from 1.7 million tons to 2.5 million tons, but that domestic production would be less than 800,000 tons per annum. The reserves of ore with at least 1 percent copper were essentially fixed in size, and the copper industry was not likely to use leaner ores. The most pessimistic conclusion was that imported copper would supply more than one-third of U.S. consumption by the early 1970s.

Actually, the industry performed much better than predicted. U.S. copper consumption in the early 1970s was about 2.3 million tons per annum, slightly below the Paley Commission's estimate. Domestic production, however, was approximately 1.6 million tons per year, and scrap copper contributed more than 900,000 tons to total consumption, eliminating the need for imports. The average grade of ore mined in the United States in 1972 was below 0.6 percent, considerably lower than a quarter-century before.[51]

Even before the Paley Commission issued its findings, the federal government had sought to encourage production of this strategic metal. The Defense Production Act, which went into effect in September 1950, offered potential producers a variety of incentives to expand output, including long-term purchase agreements; accelerated depreciation of capital equipment; and loans by the Reconstruction Finance Corporation. Between 1951 and 1953, the federal government signed agreements with six copper companies to open six new mines. Four of the projects were in Arizona, one was in Nevada, and one in Michigan. Washington gave all these companies purchase contracts paying between 22 and 27 cents per pound.[52]

The Reconstruction Finance Corporation offered loans to only two of the six ventures—$57 million for Copper Range's White Pine mine in Michigan's Upper Peninsula and $94 million for Magma's mine at San Manuel, Arizona. Neither would have come into production without RFC loans because they were too risky to attract private investment. More important, the six mining companies invested a total of $129 million of private funds in these mines. Four of the six ventures developed as truck-operated open-pit porphyry mines. The San Manuel mine began as an underground porphyry operation and eventually became the third largest underground copper mine in the world. The White Pine mine was an underground low-grade sulphide ore operation. The projected total production of the six mines was 228,000 tons, roughly 20 percent of national copper output for 1950–53. Except for the San Manuel mine, all of the subsidized mines achieved their projected output levels by 1955. San Manuel had shoring problems, which Magma finally resolved in the late 1950s.[53]

The U.S. copper industry enjoyed relative economic stability between 1957 and 1963. Prices remained between 30 and 32 cents per pound except for the recession year of 1958, when prices dipped to 26 cents. A long strike reduced production in 1959 to only 825,000 tons, but with that one exception, annual output remained between 1 million and 1.2 million tons from 1957 to 1963. The average return on investment for the six largest firms (table 7.2) remained healthy, although lower than in the early 1950s. The period 1950–63 was the first time since the early the twentieth century that the U.S. copper industry earned healthy profits during a long period of relative peace.

TABLE 7.2
Average Return on Investment for Leading U.S. Copper Companies, 1950–1963

	1950–1956	1957–1963
Phelps Dodge	19.0%	9.6%
Newmont	17.4	12.1
AMAX	16.7	15.0
Kennecott	16.6	8.4
ASARCO	13.6	6.9
Anaconda	8.0	5.0

Source: Thomas R. Navin, *Copper Mining and Management* (Tucson: University of Arizona Press, 1978), 155, 159.

The managers of the large copper firms generally were reluctant to invest their earnings in further development of domestic copper. In 1956, four of the six largest companies had no long-term debt. AMAX carried a small debt of only $14 million, while Anaconda had long-term debt of $112 million and net assets of $789 million. Anaconda's borrowing was almost entirely the result of its abortive entry into the aluminum industry. By 1962, three of the six companies still had no long-term debt, while AMAX (at $10 million) and Kennecott (at $12 million) had only modest ones. Anaconda had lowered its debt to $65 million, and its net assets stood at $972 million.

The copper giants became ultraconservative in their investment strategies. Except for Anaconda, the leading firms held large amounts of surplus funds in government securities. In 1956, for example, Phelps Dodge held $141 million, or 35 percent of its net assets, in this form, while Kennecott held $249 million, nearly 40 percent of its assets. As if they could see no point in reinvesting in copper, the large firms distributed most of their after-tax profits to their stockholders. Over the years 1957 to 1963, Kennecott paid out 86 percent of its profits as dividends, and Phelps Dodge disbursed 81 percent in the same fashion. The other four large copper companies paid out between 49 and 61 percent.[54]

Was the copper industry's dismal performance after the Second World War largely the result of ultraconservative, lackluster management? Thomas Navin has argued that, by the 1930s, the six largest

U.S. copper companies had inflexible, short-sighted leaders. They were engineers who spent their entire careers with a single firm and were skilled at cost-cutting but were not aggressive competitors. Metallurgists became the chief executive officers of the largest firms starting in the mid-1950s. Phelps Dodge was the exception, with a lawyer in control. They had experienced collapsing markets following World War I and during the Great Depression. At the end of World War II, they could only foresee a return to the conditions of the 1930s, so they were reluctant to invest in new capacity.[55]

Some contemporary observers of American business were less charitable than Navin in assessing the abilities of the copper company executives after World War II. Herbert Solow, writing in the January 1955 issue of *Fortune,* argued that the Anaconda Company suffered from managerial arterial sclerosis, controlled by aged executives who had spent their entire careers in Montana. Anaconda lacked a modern management structure and had no executive training program to develop new leadership. Anaconda's response to Solow was to hire Isaac Marcosson, a professional writer of company histories, to prepare a book praising the company's past and current management.[56]

In the cases of Anaconda and Kennecott, the low-cost, profitable copper produced by their overseas mines may have discouraged investment within the United States. Many large U.S. copper companies, however, could not explain their failure to invest in those terms. Two monuments to copper company executive failure in the postwar years stand out. First, the only significant investment in new copper mines between 1945 and 1963 took place under the federal Defense Production Act program of subsidies and price guarantees in the early 1950s. More remarkable, perhaps, is the fact that oil companies discovered *all* of the significant new porphyry copper deposits found within the United States between 1964 and 1974, a total of eight major deposits.[57]

Labor Relations since World War I

Labor relations in the copper industry were hostile and chaotic for most the period covered by this chapter. Most of the copper industry functioned without "legitimate" labor unions for much of this period and then faced rival unions claiming jurisdiction over the same groups of workers. In the late 1950s, two new entrants into this labor field, the United Steel Workers of America (USWA) and the United Automobile Workers (UAW), organized most of the workers in copper mines, mills, smelters, and fabricating plants. The attitudes of the copper companies about organized labor had changed very little since the 1910s, but federal legislation, mainly the National Labor Relations Act (1935), forced them to bargain with legitimate labor unions. The leading companies, especially in the more isolated settings in the Southwest, continued some paternalistic activities, such as providing housing for workers, but tried to disengage themselves from other forms of paternalism.

A series of violent but unsuccessful strikes against the copper companies in the 1910s had eliminated organized labor from most of the copper mining districts. The Michigan district strike of 1913–1914, the destruction of the Butte Miners' Union in 1914, and the utter defeat of the Arizona copper unions in 1917 and 1918 were the major setbacks. The International Union of Mine, Mill, and Smelter Workers (IUMMSW) in essence died during the depression years of 1920 and 1921, although it survived, in name only, into the early 1930s. Throughout the 1920s, the copper industry in Michigan, Montana, Arizona, Utah, and Nevada remained largely free of unions.[58]

In the mid-1930s, copper workers reestablished local unions and commenced bargaining under the protective umbrella of federal labor legislation. A revived IUMMSW led the organizing efforts and became a founding member of the Congress of Industrial Organizations (CIO) in 1938. For most of its history, the IUMMSW faced bitter jurisdictional disputes with various AFL and CIO unions, internal leadership struggles, and battles with both the CIO and the federal government over alleged Communist domination. It is instructive that Vernon H. Jensen wrote two books about organized labor in this industry between 1932 and 1954. The first, which analyzed the leadership problems of the IUMMSW, has more than 300 pages of text, while the second book, which focused on the bargaining history of the industry, is only 69 pages long.[59]

The revival of unions in the copper industry began in Butte in 1934, when the IUMMSW struck the Anaconda Copper Mining Company for four months and won a contract. Local unions also emerged in many Utah and Nevada mines and smelters, and these received recognition in 1936 as the result of strikes. The IUMMSW was unable, however, to organize workers at the giant Utah Copper Company operation at Bingham Canyon or the ASARCO smelter at Garfield. Scattered small locals in a half dozen states also emerged by 1936, when the IUMMSW had 15,000 dues-paying members nationwide.[60]

The IUMMSW had little success in representation elections conducted by the National Labor Relations Board (NLRB) in 1937 and 1938, winning less than half the contests it entered. The Utah Copper Company, for example, had established a company union that called for a "no-union" vote in several NLRB-supervised elections in 1937 and 1938. The workers ultimately voted for the company-sponsored union and overwhelmingly defeated the IUMMSW. Strong company unions also slowed the IUMMSW's efforts to organize Phelps Dodge's Arizona mines. The turning point in organizing was 1939, when the union's success rate in winning NLRB elections rose to 90 percent. The Michigan mines all came under contract by 1942. Because of the successful organizing drives, IUMMSW membership rose to roughly 50,000 by mid-1941 and to 95,000 in 1944. By the latter year, the union had organized most of the employees of Phelps Dodge, Kennecott, and ASARCO. The union engaged in little bargaining during the war because the National War Labor Board set wages and hours.[61]

Internal struggles over control of the IUMMSW were bitter and divisive in the late 1940s, starting with the annual convention of 1946. Reid Robinson of Butte, Montana, who was a conservative unionist when he became president in 1936, came under the control of Communists. He resigned in 1946, but his successor, Maurice Travis, was more openly sympathetic to the Communists.

The Taft-Hartley Act of 1947 required union leaders to sign a "Non-Communist" affidavit to enjoy the protection offered to unions by the federal government, but the IUMMSW refused to comply. In doing so, the union could no longer participate in representation elections and faced many employers who used the union's position as an excuse to break off bargaining. Dozens of locals opposed to the union's Communist leadership seceded in 1947 and 1948, removing thousands of members from the rolls. The dues-paying membership, which had peaked at 114,02 in 1948, fell to 68,651 in 1950. Finally, the CIO investigated Communist domination of the IUMMSW and in 1950 expelled it from the national labor federation. The CIO gave the USWA jurisdiction over workers in mining, smelting, and refining operations, while the United Automobile Workers (UAW) had jurisdiction over workers in metal fabrication plants.[62]

As late as 1956, most labor contracts in the copper industry covered only a single mine or plant and had widely varying expiration dates. In 1964, however, the IUMMSW negotiated multiplant (not companywide) agreements that would expire simultaneously in 1967, thus beginning the move toward industrywide bargaining. Between 1955 and 1966, the leadership of the Mine Mill union was under constant legal attack by the U.S. Justice Department because of alleged links with the Communist Party. In 1967, after the Justice Department stopped its prosecutions, the IUMMSW became a department of the USWA.[63]

The U.S. copper industry's worldwide competitive position thus declined substantially between the early 1920s and the early 1960s. The United States became a substantial net importer of foreign copper, while exhausting a substantial share of its copper reserves. More important, the closing of entire mining districts and large cutbacks in others have left a residue of unemployed workers and devastated communities. By the mid-1960s, copper mining had practically ended in Michigan and was barely surviving in Montana. These long-term downward trends have continued unabated since the mid-1960s. The next chapter will carry the history of the industry to the recent past.

CHAPTER 8

A Quarter Century of Adjustment and Decline

With what seems an incredible abruptness, the last five years have witnessed the utter devastation of the (American) West's oldest industry, metal mining. True, this notoriously unstable and cyclical industry has fallen before, as in the 1890s or the 1930s, but this time there seems no possibility that metal mining can recover to anything like its former status in the region.—Michael Malone in the *Pacific Historical Review* (August 1986)

The gradual decline of the U.S. copper industry in the twentieth century, which began in the 1920s, continued into the quarter century following 1963. The U.S. share of global production fell from 24 percent in 1960 to 14 percent in 1986, when Latin America produced one-quarter of the world's copper, Africa accounted for one-sixth, and the Soviet Union equalled U.S. production. Although world copper production doubled between 1960 and 1986, American output increased by only 20 percent over the same span. Employment in mining, smelting, refining, and fabricating, 150,000 in 1950, fell to 77,300 in 1972 and to only 34,400 in 1986.[1]

Adding to their troubles, the major U.S. copper companies lost most of their overseas mines during these years. Over the span of 1965–74, the governments of Mexico, Peru, Zambia, and Chile asserted national control over the foreign-owned copper mines within their borders. The three major U.S. firms affected—Anaconda, Kennecott, and AMAX—lost nearly half their global production capacity.[2]

In recent decades, most of the major U.S. copper companies have no longer focussed their efforts on copper and its precious-metals byproducts. After 1964, the six largest copper companies diversified into lumber, cement, coal, lead, zinc, nickel, and aluminum. The distinction between the petroleum industry and the copper industry became blurred during this era as well. The oil producers also made U.S. copper companies targets for corporate takeovers. By the early 1980s, more than a dozen copper producers—including Calumet and Hecla, Anaconda, and Kennecott—had become subsidiaries of oil companies.[3]

Market conditions were more volatile after 1963 than in any previous period. A sharp increase in demand resulting from the Vietnam War produced temporary price increases and government price controls. An industrywide strike in 1967 further disrupted normal operations, as did the loss of overseas mine properties between 1965 and 1974 mentioned above. During the early 1980s, the U.S. copper industry faced rising costs of production and sharply falling copper prices. The large producers either scaled back their operations or entirely shut down because of these conditions.

The Loss of Foreign Properties

Mexico started the international movement toward indigenous control of mining by enacting legislation in 1961 granting substantial tax advantages to mining companies with majority ownership of stock by Mexican nationals. AMAX immediately moved to sell a majority interest in its properties to Mexicans, and by 1965 it had liquidated all its Mexican investments. Anaconda and ASARCO ignored the law until 1965, when the Mexican government forced the issue. Although U.S. companies did not receive what they thought was a fair price for their properties, they at least received cash and were not expropriated outright. In 1980 the Mexican government forced all mining companies to become mixed public-private enterprises.[4]

The establishment of the nation of Zambia (formerly Northern Rhodesia) in 1964, with political control by its black majority, began the end of foreign dominance of that country's copper mines. Two firms influenced by American capital, Rhodesian Selection Trust (RST) and Rhodesian Anglo-American Company, had developed the major copper deposits in the late 1920s and 1930s. The American Metal Company, later AMAX, controlled 42 percent of RST stock and operated its mines and smelters. Following World War II, American capital withdrew from Rhodesian Anglo-American, giving control to British and South African investors.

The new government of Zambia launched a program in the late 1960s to replace British and American skilled tradesmen and managers with Zambian nationals. In 1970 the new nation's president, Kenneth Kaunda, "invited" the copper companies operating in Zambia to sell 51 percent of their stock to a new government agency, Mindeco. They received government bonds as payment but also earned generous management fees and sales commissions. The Zambian government bought out AMAX's remaining interest in 1974, paid off the bonds, and ended AMAX's management and marketing agreements. AMAX received generous compensation for its Zambian properties.[5]

Similarly, the government of Peru nationalized its copper mines but paid the foreign companies compensation for their losses. Juan Velasco Alvarado became president of Peru following a military coup in 1968, and in September 1969 he ordered companies holding mining concessions to submit plans to develop the properties. Velasco's main target was the Cerro de Pasco Corporation, which had been the dominant producer in Peru in the 1910s but which had failed to modernize its underground mines since World War II. Instead, the firm developed an open-pit mine in the late 1950s, enlarged its smelting plants, and operated vast agricultural estates. After Peru nationalized Cerro de Pasco's agricultural estates in 1969, its owners offered to sell their properties to the government. In 1972, Peru offered $12 million for Cerro's properties, which had a book value of $145 million and a market value of about $250 million. When subsequent negotiations failed, the Peruvian government expropriated Cerro's holdings on 1 January 1974 and later paid $58 million as compensation.[6]

Peru's other major foreign-owned copper producer, the Southern Peru Copper Corporation (SPCC), was a wholly owned subsidiary of ASARCO. In 1954 the Peruvian government signed a contract with the SPCC establishing the firm's rights and obligations in developing its property at Toquepala. Work began there in 1956, and after the mine reached full production in 1960, it accounted for more than half of

Peru's copper output over the next two decades. In the mid-1960s, the Velasco government also negotiated an agreement with the SPCC to develop the 400-million-ton porphyry deposit at Cuajone, just northwest of Toquepala.

After the Peruvian government issued a new mining law in June 1971, which guaranteed that foreign mine operators would reinvest most of their profits in Peru, the SPCC announced that it would spend $48 million to develop the Cuajone deposits. The SPCC opened the Cuajone mine, but the Peruvian government owned the refinery that treated the ore. Once the mine came on line in 1977, it produced 40 percent of Peru's copper.[7]

The nationalization and expropriation of American-owned mines in Chile between 1969 and 1971 permanently altered the structure of the global copper industry. Chilean attacks on the American copper companies' domination of the nation's mining industry developed as early as the 1920s but became more frequent and more strident in the 1950s. The basic criticism was that the companies earned huge profits, which went abroad to their American stockholders. In addition, most of their spending on supplies, machinery, and equipment went to firms based in the United States. Critics also accused Anaconda and Kennecott of restricting output at their mines and failing to develop additional deposits, therefore depriving Chile of income needed for economic development. The U.S. copper companies exploited Chile's most important natural resource with little benefit to the Chilean economy.[8]

Eduardo Frei won Chile's 1964 presidential election by advocating the "Chileanization" of the copper mining industry. Much to Frei's surprise, Kennecott offered to sell 51 percent of its equity in El Teniente to the government. Kennecott accepted state ownership as inevitable and structured its presence in Chile to minimize its future losses. The Chilean government, the Export-Import Bank, and private bankers paid for expanding production, with Kennecott assuming no financial liability. The firm retained a 49 percent equity in a greatly expanded El Teniente mine and, with more favorable tax terms, earned as much profit under the new arrangement as under the old. The Chilean government agreed to compensate Kennecott for its loss of equity with payments of $80 million over twelve years.[9]

The contrasting experiences of Kennecott and Anaconda in Chile are instructive. Anaconda tried to maintain its corporate sovereignty in Chile by refusing to sell equity in its mines. It launched a new company to develop the Exotica mine, located near Chuquicamata, but would not sell majority control of the venture to the government of Chile. In 1966, Chile's Congress established a new government corporation, CODELCO, to serve as a partner in joint ventures with foreign firms. Frei continued to pressure Anaconda for state participation in the Chuquicamata and El Salvador mines. Anaconda avoided a further confrontation until 1969, largely because the company's politically astute president, Charles M. Brinkerhoff, was a close friend of President Frei. Anaconda's strategy worked until Brinkerhoff retired in 1968.

In 1969, Frei forced Anaconda to sell 51 percent of the equity in its properties to the Chilean government in return for $175 million in twelve-year government bonds yielding 6 percent interest. Anaconda also granted Chile an option to buy the remaining 49 percent after 31 December 1972, with payments in interest-bearing government bonds paid over twelve years. CODELCO would have a majority on the boards of directors of both mines, but Anaconda would control day-to-day operations for at least three years and would receive a management fee of 1 percent of gross sales.[10]

The Chilean presidential election of 1970 made these agreements moot. Salvador Allende, the leader of a left coalition, became president on 24 October 1970. On 11 July 1971, Chile's legislature passed a constitutional amendment bringing the mines under state control. Allende agreed to compensate the American companies, but Chile would not assume any debts unless the foreign companies had spent the borrowed funds wisely. Chile would also deduct the value of "excess profits" from the money owed to the U.S. investors. On 11 October 1971 the government of Chile announced that Anaconda and Kennecott owed Chile $378 million net. Anaconda and Kennecott countered by using the courts in the United States and Europe to seize Chilean government assets. Allende retaliated by cutting off copper shipments to Anaconda, forcing the firm to buy the supplies it needed for its manufacturing operations on the open market.

A military coup in Chile on 11 September 1973 ended Allende's regime and his life. The new military regime agreed to compensate Anaconda and Kennecott with long-term government bonds, with the valuation based on book value, less than half market value. The impacts of the loss of the Chilean mines on Kennecott and Anaconda were drastically different. Kennecott lost 11 percent of its global copper production but little of its equity once bond payments resumed. Anaconda, however, lost 75 percent of its copper production and one-third of its net worth. These losses, combined with failed efforts to expand mining capacity in Arizona in the early 1970s, created a severe financial crisis for Anaconda and helped make the company a target for a corporate takeover.[11]

Corporate Takeovers and Diversification

The most striking development in the U.S. copper industry from the early 1960s to the early 1980s was the purchase of many large producers by oil companies. Between 1963 and 1982, nine companies lost their independence, including Calumet and Hecla, Copper Range, Anaconda, and Kennecott. Takeovers occurred because oil companies wanted to diversify, had vast amounts of cash to invest, and believed that copper mining was close enough to their existing activities to make the risks acceptable. Some copper companies became targets because of their weakened financial condition and their failure to diversify profitably. With few exceptions, the oil companies regretted these acquisitions.[12]

The Calumet and Hecla Mining Company, which had successfully diversified into lumber and plastics in the 1950s, prospered into the next decade. In early 1968, the Universal Oil Products Company (UOP) offered $123.5 million for Calumet and Hecla's stock, and the stockholders accepted. A United Steel Workers strike had shut down the C & H mines since August 1967, and when the union refused to end the strike, UOP permanently closed the mines in March 1968. The other significant Michigan producer, the Copper Range Company, operated the White Pine mine and was a marginally profitable producer. In 1975, AMAX and Copper Range nearly consummated a "friendly" merger, but the U.S. Justice Department blocked the agreement. The following year, the Louisiana Land & Exploration Company, an oil and land development firm, purchased Copper Range, marking the end of copper mining in Michigan by independent firms.[13]

The Anaconda Copper Mining Company's financial health became critical in the early 1970s because of the loss of its Chilean mines, the rapid depletion of the Butte deposits, and an ill-conceived investment in Arizona. In 1971, Anaconda began developing a large, low-grade deposit at Twin Buttes, Arizona, south of Tucson, with the Banner Mining Company as a partner. Twin Buttes had both sulphide and oxide deposits, but Anaconda processed only the sulphide ore and stockpiled the oxide ore because it lacked the capital to fully develop the property.

In 1973, AMAX bought out Banner and agreed to provide enough capital to fully develop the deposits. AMAX and Anaconda were to operate Twin Buttes as a 50-50 joint venture via a new corporation, the Anamax Mining Company. AMAX honored its commitment to Twin Buttes, expanded production there, and took over its management, but neither firm earned a profit there.

Anaconda's profits in the early 1970s came increasingly from its brass and aluminum businesses rather than from copper. In 1975 the Crane Company, a brass manufacturer, attempted a hostile takeover of Anaconda, which the potential victim stymied by buying out one of Crane's competitors in the brass industry. Crane could not vote its Anaconda stock without incurring the wrath of the U.S. Justice Department for antitrust violations. To end the stalemate, the Atlantic Richfield Corporation (ARCO), Crane's parent company, acquired Anaconda in 1977. ARCO quickly lost $750 million in Montana and ended mining operations there in 1983. Montana Resources, Inc. (MRI) reopened Anaconda's Continental Pit in Butte east of the Berkeley Pit and the nearby Weed Concentrator in the late 1980s, and shipped copper concentrates to Japan for smelting.[14]

The Phelps Dodge Corporation has fared better than most of its competitors because it has suffered no substantial losses of capital overseas. By the mid-1970s, Phelps Dodge was the second largest copper producer among U.S. companies, slightly behind Kennecott but twice the size of Anaconda. Before Phelps Dodge celebrated its centennial in 1981, mining operations had ceased at Bisbee with the closing of the Lavender Pit in 1974 and the end of underground mining at the Copper Queen the following year. The firm also permanently shut down its huge open-pit mine at Ajo in 1984 in the midst of a strike. In 1969, the firm redeveloped the low-grade deposits at Tyrone, New Mexico, which the company had abandoned in 1921. To further boost its reserves, Phelps Dodge also bought the Chino mine at Santa Rita, New Mexico, from Standard Oil of Ohio (SOHIO) in 1986. Through all these changes, the Morenci mine, the giant open-pit operation that went into production in 1942, has remained Phelps Dodge's mainstay.

By the late 1970s, Phelps Dodge's Arizona smelters (Ajo, Douglas, and Morenci) were not meeting federal standards for air quality. In February 1981, Phelps Dodge reached a consent agreement with the federal Environmental Protection Agency (EPA) giving the company until 1 January 1985 to comply with federal standards. Phelps Dodge, however, concluded that it could not economically meet the standards and closed its smelters at Ajo and Morenci in 1984. The Douglas smelter received several reprieves before closing for good in January 1987. Meanwhile, Phelps Dodge built a smelter at Hidalgo, New Mexico, completed in 1976 and equipped with a new technology, the Finnish flash smelting (Outokumbu) process. This technology burns the sulphur in the copper concentrates as fuel, permitting drastic reductions in pollution. The Hidalgo smelter eventually replaced all the Arizona smelters.[15]

The Kennecott Copper Corporation was the copper producer most committed to the exploration and development of new domestic mines. Kennecott spent more than $100 million on exploration in the early 1960s alone and focussed on North America. There was little payout, however, because the firm adopted a conservative approach to this work. Kennecott instructed its geologists to limit exploratory work to depths of 1,500 feet, since the company did not want to develop underground mines. Their geologists were also too conservative in assessing the economic viability of low-grade deposits. Kennecott thought that the Esperanza and Mineral Park deposits, both in Arizona, were too poor to develop, but the Duval Corporation, a newcomer, profitably mined them.[16]

Kennecott's American mines remained profitable in the mid-

1960s, and the company had $200 million in excess cash but decided against spending those funds on exploration or improving the efficiency of its U.S. operations. Instead, Kennecott looked to diversify and in January 1968 bought the Peabody Coal Company, the largest independent coal producer in the United States. Kennecott paid about $622 million for Peabody, which independent analysts thought was worth only $475 million. The Federal Trade Commission (FTC), however, brought suit, ruled that the merger restrained trade, and ordered Kennecott to divest itself of Peabody Coal. The company finally sold Peabody Coal for $1.2 billion in 1977 to a holding company. Later that year, Kennecott spent half its cash, $567 million, to buy the Carborundum Company, a decision that financial analysts condemned as ill-advised.[17]

Kennecott's large cash reserves in the late 1970s made it an attractive target for takeover bids. In 1981, SOHIO, a wholly owned subsidiary of British Petroleum, bought Kennecott for $1.7 billion. SOHIO began a $400 million modernization program at Bingham Canyon, intending to concentrate its copper mining efforts at a single site. In 1984, SOHIO permanently closed the Nevada Consolidated mines, and in 1986, it sold its interest in the Chino mine to Phelps Dodge for $93 million and its Ray, Arizona, mine to ASARCO for $72 million.

After losing $600 million in copper mining, SOHIO sold Kennecott in 1988 to RTZ (Río Tinto Zinc) Minerals, Ltd., a British-owned multinational mining firm, for $3.7 billion, completing the dismemberment of Kennecott. RTZ renamed its new property the Kennecott Utah Copper Company and committed an additional $227 million for further improvements at Bingham Canyon. It is ironic that, unlike the Kennecott executives of the 1960s, the corporate leaders of SOHIO and RTZ saw the wisdom of investing in Bingham Canyon. RTZ returned that mine to its past glory as one of the largest, lowest-cost copper mines in the world.[18]

ASARCO continued to integrate vertically by buying additional mining capacity and remained the largest copper smelter in the world. Much of ASARCO's growth was overseas. The firm held on to its Mount Isa mines in Australia while increasing its investment in Peru with the development of the Cuajone mine from 1971 to 1977. By the mid-1970s, half of ASARCO's profits came from mining and most of its profits originated overseas. The company's first large U.S. mine, the Silver Bell (1954) in Arizona, stopped working in 1982 but continued as a leaching operation. ASARCO opened the San Xavier mine in 1974 near Mission, Arizona, and bought two small mines in Pima County, Arizona, the Pima and the Eisenhower, in the early 1980s. The company also bought the Ray mine and remained a major force in mining as well as smelting.[19]

The Newmont Mining Corporation, which began as a holding company with a minority interest in the Magma Copper Company, gained control of all of Magma's stock in 1969. Magma bought a deposit (the Kalamazoo mine) next to its San Manuel mine in 1971, doubling its capacity. During the 1970s, half of Newmont's profits came from the Southern Peru Copper Company and Magma, but when Magma had large losses in the 1980s, Newmont spun it off as an independent company in 1987 while giving Magma the Pinto Valley Copper Corporation, another small Arizona copper company. Magma invested $300 million in new facilities in the late 1980s and in 1990 reopened its historic Superior, Arizona, mine.[20]

AMAX was one of the six largest U.S. copper companies after 1965, but much of its metal came from mines in Mexico, Peru, and Africa. Facing the eventual loss of its foreign mines, AMAX joined Anaconda in developing the Twin Buttes mine in Arizona in 1971 and sank large sums into this unprofitable venture. In the early 1980s, AMAX closed its molybdenum mine at Climax, Colorado, and sold most of its U.S. mines, including Twin Buttes, to Cyprus Minerals. By 1990 AMAX was no longer producing copper.[21]

New copper producers emerged in the 1960s and later. The Duval Corporation, which had begun as the Duval Texas Sulphur Company in 1926, diversified into copper after 1959, when it purchased the Esperanza orebody, south of Tucson, Arizona, and the Mineral Park orebody near Kingman, Arizona. Under Vietnam War–era legislation to encourage the development of new copper mines, Duval received an $83 million loan from the General Services Administration to develop its low-grade (0.28 percent copper) Sierrita deposit, located near its Esperanza mine. When it opened in 1970, Sierrita had an annual capacity of about 80,000 tons of metallic copper. By 1974, when Duval was operating only one mine, it was the fifth largest

TABLE 8.1

Annual Capacity of the Largest U.S. Copper Mines, By Firm, 1988

	Tons
Phelps Dodge	359,000
RTZ Minerals	200,000
Cyprus Minerals	182,000
Magma	172,000
ASARCO	149,000
Montana Resources	89,000
Copper Range	51,000

Source: George H. Hildebrand and Garth L. Mangum, *Capital and Labor in American Copper, 1845–1990: Linkages between Product and Labor Markets* (Cambridge, Mass.: Harvard University Press, 1992), 170.

copper mining company in the United States. The Pennzoil Company, which had bought a controlling interest in Duval in 1968, sold its copper properties to Cyprus Minerals in 1986, ending Duval's brief involvement in the copper industry.[22]

Cyprus Minerals Corporation was another late entrant into copper mining in the United States. Seeley Mudd and Philip Wiseman, who had helped develop the porphyry mine at Ray, Arizona, established the Cyprus Mines Corporation in 1914 to develop low-grade deposits on the island of Cyprus. The company began copper production in the U.S. after buying the Pima, Arizona, orebody in 1955 and the Bagdad, Arizona, mine in 1956. Cyprus increased Bagdad's annual capacity from 20,000 tons of metallic copper to 70,000. Standard Oil Company of Indiana (AMOCO) bought Cyprus in 1979 but spun it off in 1985. In the late 1980s, Cyprus Minerals purchased an impressive set of copper mines: Twin Buttes from ANAMAX; the Esperanza-Sierrita complex in Arizona from Duval; the Lakeshore (Casa Grande) mine in Arizona; and the Inspiration Consolidated Copper Company's Arizona properties.[23]

Table 8.1, which lists the annual capacity in 1988 of the largest mines owned by the major corporations, illustrates the overall structure of the U.S. copper industry in the late 1980s. The contrast with the mid-1960s is striking. Three of the top six copper companies in 1965 had disappeared—Anaconda, AMAX, and Kennecott—although RTZ Minerals was producing all of its copper from Kennecott's Bingham Canyon mine.

The decline of the American copper industry in the quarter-century between the early 1960s and the late 1980s continued the trends that had been in existence since the 1920s. The decisions made by the top management of the major American copper companies only speeded that decline. They decided that copper mining within the United States would never again yield healthy profits, so they failed to search for new deposits or invest in the mines already working. Instead, they invested in their overseas properties and in a series of failed efforts to diversify into other industries. The lone exception was Phelps Dodge, which became the leading American copper producer in the late 1980s.

American Copper in the Global Context

Since World War II, the American copper industry has been an increasingly less important segment of an expanding global copper industry. By the mid-1980s, socialist countries produced 24 percent of world copper output, while state-owned mining enterprises in the less-developed countries accounted for another 30 percent. In 1984, private multinational companies controlled only 7 percent of the copper produced in less-developed countries.[24]

The growth in worldwide demand for copper slowed noticeably after the early 1970s. The rate of growth in copper consumption in the market economies slowed from an annual rate of 4.8 percent from 1963 to 1972 to only 1.2 percent per annum from 1973 to 1984. Growth in the first period reflected expanding industrial production and the demands of the Vietnam War. The growth of industrial production worldwide slowed between 1973 and 1984, reducing the growth in demand for traditional industrial metals. The consumption of steel, lead, tin, and zinc actually fell in these years. More important, growth took place in "high-technology" industries such as computers, which

TABLE 8.2
Copper Production Costs for Major Market Economies

	1975	1984
United States	61.6[a]	78.1[a]
Canada	28.4	56.1
Peru	51.1	56.8
Chile	47.2	48.8
Zaire	55.1	45.2
Mexico	27.3	37.9

Source: George H. Hildebrand and Garth L. Mangum, *Capital and Labor in American Copper, 1845–1990: Linkages between Product and Labor Markets* (Cambridge, Mass.: Harvard University Press, 1992), 197.
[a]Cents per pound in nominal U.S. currency.

used a wide range of new industrial materials, including polymers, silicon, optic fibers, and metals such as platinum, titanium, beryllium, and lithium instead of copper.[25]

The growth in the production of copper worldwide slowed to an annual rate of only 1.5 percent over the period 1970–86. The areas of most rapid growth were Eastern Europe (7.3 percent per annum), Asia (6.3 percent), and Latin America (5.1 percent). Production from Africa as a whole fell slightly between 1970 and 1986 because a 25 percent drop in output from Zambia canceled out increased production from Zaire. U.S. mine production actually fell by 1 percent per annum during 1970–86, a decline of about 20 percent.[26] Most of the decline in U.S. production took place after 1980, when prices moved downward. By then, American mines had become high-cost producers (see table 8.2).

Adding to economic instability, American copper prices were more volatile during the quarter-century beginning in 1963 than in any similar period over the last two centuries. The Vietnam War pushed prices upward from 1965 through 1970, when they reached 71 cents per pound. Prices climbed to 71 cents a pound in 1970, fell to 54 cents in 1971–72, rose sharply to $1.03 in 1974, and fell precipitously to 62 cents in 1975. When prices declined again beginning in 1980 and costs spiralled upward, the major U.S. copper companies suffered huge financial losses. In 1982 and 1983, ARCO (Anaconda), ASARCO, Kennecott, Phelps Dodge, and the smaller operators all reported losses. Anaconda's high costs of production—twice the market price of copper—forced ARCO to close its Montana mines permanently in 1983. The other large companies temporarily shut down most of their mines for long stretches during 1982 and 1983. The industry subsequently benefited from higher copper prices in the late 1980s, combined with reduced costs, especially for labor.[27]

Labor Relations, 1967 to 1986

Labor relations within the copper industry became noticeably more contentious after the merger of the IUMMSW and the United Steel Workers of America (USWA) unions in January 1967. The Steel Workers' leadership created the Nonferrous Industry Conference (NIC) to coordinate and approve demands from individual unions and locals within the nonferrous industries. The NIC was a coalition of twenty-six national unions which set common strike goals, negotiated jointly, and followed a common strike strategy. The NIC set the following goals for 1967 bargaining: companywide master agreements covering all plants and including subsidiaries; common expiration dates for agreements; and uniform wages and benefits.

The copper companies firmly resisted this attempt to change the industry's bargaining structure, and a strike began on 15 July 1967. With the strike in its eighth month, President Lyndon Johnson called the parties to Washington in early March 1968 and asked them to stay until they resolved the strike. Over the course of a month, the major companies settled with the unions but did not grant companywide agreements, much less industrywide uniform wages. The fact that the NIC held together for nine months was the only "victory" the USWA could claim. The resolve shown during the 1967–68 strike made the USWA-led coalition more formidable in subsequent disputes.[28]

Starting with the 1971 negotiations, the union coalition adopted a new approach. Rather than trying to reach a settlement with the entire nonferrous industry, the union coalition targeted the basic copper industry alone, excluding the fabrication plants. The coalition focussed on the ten largest companies and adopted a pattern-bargaining approach. The unions would reach the best possible agreement with a single company and then demand the same contract from the others. The companies that settled could then capture part of the market held by the firms that the unions closed down through strikes. The industry suffered limited walkouts, usually lasting about a month, in July of 1971, 1974, and 1977, with a longer shutdown in 1980 extending more than three months. The union coalition won Cost of Living Adjustments (COLAs) in the 1971 settlements and spent the rest of the decade successfully defending that gain.[29]

The American copper industry faced a classic "scissors crisis" starting in 1980, with copper prices dropping sharply while wages simultaneously increased. Nominal copper prices, which had averaged \$1.09 per pound in 1980, dropped to \$0.80 for 1983 and then averaged \$0.70 from 1984 to 1986. Wages rates, however, increased at an annual rate of roughly 14 percent from 1980 to 1986, in part the result of COLA provisions in the industry's labor contracts. In this grim economic setting, the United Steel Workers and Kennecott announced agreement on a new three-year contract in mid-April 1983, ten weeks before the existing contract expired. The agreement froze basic wage rates but continued COLAs, which the contract rolled into base wages annually. Kennecott agreed to these provisions because the unions allowed the company to drop an employment security plan it had granted in the early 1960s. The union coalition eventually imposed this agreement on the rest of the industry, with one major exception. The Phelps Dodge Corporation reached an impasse with the unions but decided to operate despite a strike.[30]

The Phelps Dodge management offered a series of radical proposals to the unions: the end of the COLA system and the introduction of a two-tier wage structure, with new hires earning much less than established employees. The union coalition refused to accept any variance from the Kennecott pattern, so the strike that began on 1 July 1983 was inevitable. Phelps Dodge, however, continued to operate its facilities with managers, nonunion workers, union members who refused to strike, and replacement workers. The unions continued the struggle until December 1984, but by then they had lost decertification elections and Phelps Dodge had become a nonunion company again.[31]

With the copper industry remaining in a depressed condition, the companies and the unions alike entered the 1986 negotiations with sharply different attitudes than in 1983. The major companies and the unions settled for a 20 percent across-the-board pay cut and an end to the COLA system. By the late 1980s, the unions were in many respects further away from the goal of industrywide uniform wages and benefits than they had been in 1967. Several important mines, such as Bingham Canyon, were nonunion, while Phelps Dodge had both union and nonunion plants. Those companies that were unionized had a bewildering variety of bargaining units, basic pay rates, bonus pay schemes, and work rules in their contracts, which expired at widely different times. This was not the contract structure the unions had wanted for a quarter of a century, but it accurately reflected the chaotic conditions of the American copper industry as it left the troubled 1980s.

A Long-Range Perspective

The declining fortunes of the American copper industry since the mid-1960s has obscured the enormous achievements of the previous century or more. The opening of the Michigan copper district in the mid-1840s, with its remarkable deposits of native copper, marked the beginning of an American copper industry with the capacity to supply the nation's domestic requirements and to serve overseas markets as well.

To be sure, dozens of small mines had produced significant quantities of the red metal before the 1840s. Two of the earliest to operate over an extended period—the Simsbury mine in Simsbury, Connecticut, and the Schuyler mine near Newark, New Jersey—went into production around 1715. American copper mining virtually ceased between roughly 1780 and the early 1830s, although copper smelters and rolling mills operated from 1801 on. A substantial revival of mining occurred in the 1830s and early 1840s in the eastern United States, but the opening of the Michigan district completely overshadowed this resurgence.

The Michigan mines supplied one-third of American demand in 1855 and three-quarters of U.S. consumption by 1860. The Michigan copper district then dominated the national copper industry until the early 1880s, accounting for more than 80 percent of U.S. output over the period 1847 to 1880. The Michigan producers protected their position in the domestic market and took full advantage of their market power. They enjoyed substantial tariff protection from 1869 to 1890. The Lake Superior producers also created an effective selling "pool" to limit production, control exports, and stabilize domestic copper prices. The Lake "pool" began in 1870 and was effective until the mid-1880s, when independent mines in Montana and Arizona began to flood the market. The largest producers in Michigan, Montana, and Arizona then joined the Secrétan Syndicate, which unsuccessfully attempted to corner the world copper supply between 1887 and 1889. By then the U.S. copper industry had become a major global producer. Its share of world copper production, only 17 percent in 1880, had reached 43 percent by 1890 and exceeded 50 percent in 1894.

Starting around 1880, the Michigan copper industry began an era of relative decline. The Lake Superior producers suffered from increasing costs as they drove to greater depths to uncover narrower veins of copper-bearing rock with declining copper content. New mining technologies, including electric haulage and the one-man drill, slowed but did not reverse rising costs of production. The Lake Superior mining district, which had produced 84 percent of American copper in 1880, accounted for only 13 percent in 1920.

The Montana copper industry, with mines at Butte and smelters and reduction works at Anaconda and Great Falls, enjoyed a colorful and meteoric entrance into the national copper industry. Most of Montana's rich copper mines, including the fabulous Anaconda, had begun as silver mines. Montana's share of U.S. copper production skyrocketed from a mere 2 percent in 1880 to 41 percent in 1885 and then peaked at 51 percent in 1895. Montana's copper industry expanded through the end of World War I, but with the appearance of new mines in Arizona and elsewhere, its position in the national copper industry deteriorated rapidly. Although copper output continued to grow, Montana's share of American production fell from 36 percent in 1905 to only 18 percent in 1916.

The history of the Montana copper district has been distinct from that of the Michigan mining industry in several respects. The capitalists who owned the Michigan copper mines lived in Boston, New York, and other eastern cities, but the men who developed the Montana mines were from Montana or other western mining states and did not become absentee owners. Montana's "copper kings"—William Andrews Clark, Marcus Daly, and F. Augustus Heinze—fought for control of Montana's copper resources, along with its government, courts, and newspapers. But the "wars of the copper kings" came to an end in 1910, when the giant Amalgamated Copper Company gained monopoly control over the Montana copper industry. Amalgamated, created in 1899 with Anaconda as its base, eventually transformed itself into the Anaconda Copper Mining Company in 1915.

The Arizona copper district, retarded by hostile Apaches and the lack of railroad connections to the outside world, developed slowly until the late 1870s. As happened in Montana, precious metal prospecting and mining served as a prelude for the discovery and exploitation of Arizona's copper deposits. Unlike the Montana experience, most of the capital needed to develop the Arizona mines came from outside of the west and in at least one case from overseas. More important, the original investors were absentee owners who relied on professional managers to run their properties. The major deposits were not located at a single site, as was the case in Montana, but in four discrete districts—Clifton, Bisbee, Globe, and Jerome.

Arizona had already emerged as a substantial copper producer in 1880 but accounted for only 5 percent of American output at the time. With the rapid development of major new mines, particularly the Copper Queen, Arizona's share of U.S. production stood at 14 percent in 1885 and reached one-fifth in 1900. The territory leaped into second place (behind Montana) in 1905 and then became the leading producer in 1910. Ten years later, Arizona accounted for 46 percent of American copper output.

The successful exploitation of low-grade porphyry deposits starting in the early twentieth century revolutionized copper mining throughout the West. In 1905, at the Utah Copper Company mine at Bingham Canyon, Utah, Daniel C. Jackling proved the economic feasibility of mining low-grade deposits through open-pit mining methods and then successfully concentrating the ores. Additional porphyry mines subsequently opened at Morenci, Arizona (1907); Ely, Nevada (1908); Santa Rita, New Mexico (1911); and at Miami (1911), Ray (1911), Inspiration (1915), and New Cornelia (1917), all in Arizona. In 1918 these eight porphyry mines already accounted for 35 percent of American copper output.

The largest U.S. copper companies opened new mines in the less developed world in the early twentieth century. The overseas properties provided them with cheap copper while protecting them against future depletion of their American copper reserves. Starting in the 1890s, Anaconda, ASARCO, and Phelps Dodge operated mines in Mexico. Kennecott and Anaconda opened mines in Chile starting in 1906, and ASARCO first invested in Peruvian copper properties in 1921. The Newmont Mining Corporation and AMAX then developed copper mines in South Africa, South-West Africa, and Northern Rhodesia in the late 1920s. Investments in foreign mines increased copper supplies for the American companies and expanded their profits as well. The major firms became multinational enterprises increasingly dependent upon cheap copper from their overseas mines. The cost of this dependence became clear in the late 1960s when the American companies lost their overseas properties via nationalization.

In the 1920s the American copper industry began a period of long-term decline compared with the copper industries in the rest of the world, a decline that has continued unabated into the 1990s. In some respects, developments that the American producers could not control had the greatest negative effects on them. The growth in the demand for copper within the United States slowed noticeably and permanently after the 1920s, while increased supplies of domestic scrap copper further dampened the demand for new copper. The costs of mining rose sharply, reflecting the lower quality of copper deposits and increased costs of machinery and labor. New foreign mines came into production in the 1910s and 1920s, flooding global markets with low-cost copper.

Since the early 1920s, the U.S. copper producers have also contributed to their worsening position in the world copper industry through their conservatism and poor decisions. They steadfastly refused to invest their profits in their existing mines to improve productivity and reduce costs. They also failed to invest in exploration or in the development of new deposits. The six new mines that opened in the 1950s as a result of the Defense Production Act of 1950 were the only exception to this dismal performance. The major U.S. copper companies either distributed their earnings to stockholders as dividends or invested in diversification. The largest firms had unimaginative, conservative leaders who preferred to maintain corporate profits simply by increasing the production of their foreign subsidiaries. The loss of virtually all of their overseas properties between 1967 and 1971 negated that strategy and forced the U.S. companies to face harsh new economic realities.

During the two decades starting in 1963, American and foreign oil companies gained control of most of the U.S. copper industry. With few exceptions, the new owners failed to earn profits in the copper industry and ultimately divested themselves of most of their copper properties. Ironically, some of the oil companies made major investments in existing mines, improved productivity, and lowered costs, and as a result earned substantial profits. The RTZ Minerals, Ltd., owner of the Utah Copper Company mine at Bingham Canyon, Utah, was the most notable example of a foreign corporation willing to make the investment needed to reduce costs and become competitive.

The Copper Industry's Legacy and Its Future

The most visible legacy of the American copper industry is the collection of abandoned mine, mill, and smelter sites and the ghost towns and depressed areas that the industry has left behind. Along the spine of Michigan's Keweenaw Peninsula, for example, are the remains and ruins of perhaps two dozen ghost towns from the copper era. The decline of Calumet, Michigan, however, is more representative of the fate of cites and towns in abandoned copper districts. Calumet and its adjacent communities had a peak population of roughly 35,000 at the 1910 U.S. Census. By 1990 the population had fallen below 2,000. The once-thriving center of Michigan's copper district, however, will not become a ghost town but instead will be the cornerstone of a new national park celebrating the district's mining heritage. Despite the near-total collapse of its economic base, the region has survived, in part because of the economic impact of Michigan Technological University, a state-supported college originally established in Houghton in 1885 as the Michigan Mining School.

The copper industry's economic legacy in Montana is less stark than in Michigan. The three principal centers of the industry—Butte, Anaconda, and Great Falls—have lost population since the early twentieth century and have suffered economic dislocation, but they have not become ghost towns. Butte's population, for example, which peaked at 42,000 in 1920, stood at 35,000 in 1990, hardly a precipitous decline. The relatively healthy condition of the Montana mining communities reflects the recent timing of the closures, with the smelter at Anaconda shutting down in 1981 and mining ending at Butte in 1983. The reopening in 1986 of the Continental Pit in Butte, along with the Weed Concentrator, both operated by Montana Resources, Inc., have partially blunted the employment impact of the Anaconda closure.

Arizona copper towns have also fared moderately well compared to their Michigan counterparts, because copper mining continues to provide employment in parts of the state. Phelps Dodge continues to employ a substantial workforce at Morenci, as does Magma at Superior and ASARCO at Ray and elsewhere. The end of mining at Bisbee in the 1980s caused economic distress there, but Bisbee has rebounded as a magnet for artists and as a tourist destination. Similarly, Jerome nearly became a genuine ghost town in the 1960s but has now revived as a tourist attraction.

The copper industry has also left permanent marks on the environment near its mines, mills, and especially its smelters. The industry has wrought permanent changes in the physical landscape in and near its production sites. Lode mining has had less serious effects on the environment and has produced less profound changes in the landscape than open-pit mining, but neither have been as destructive as hydraulic mining, commonly used in precious-metal mining. Environmental damage from copper smelters depends to a large degree on the ores smelted and the processes employed. Smelters treating the native copper of the Lake Superior district produced little damage from air pollution because the rock contains few harmful chemicals. The sulphide and carbonate ores of the West, however, typically contain, in addition to sulphur, heavy metals such as arsenic, cadmium, and lead, which may remain in the soil for generations after the smelter has closed.[32]

The neighbors of copper smelters, especially farmers and ranchers, have long recognized the damage these facilities do to their crops and livestock and have occasionally brought suit against the operators. The City of Butte passed smoke ordinances in 1890 to control this industry, but relief came only with the closure of the smelters beginning in 1893. Smelting moved to Great Falls and to Anaconda, where ranchers brought legal actions to reduce pollution starting in 1905. Similarly, the smelters at Ducktown, Tennessee, were the subject of lengthy litigation involving local complainants, the State of Georgia, and the federal government. From 1904 to 1906, farmers in the Salt Lake Valley in Utah brought successful lawsuits against smelters that had damaged their crops, causing the offending plants to shut down permanently.[33]

Besides damaging the environment through water and air pollution, the copper industry has also created permanent changes in the landscape. Ground subsidence (cave-ins) at the United Verde mine undermined and eventually destroyed a large section of downtown Jerome, Arizona. The piling of "poor rock" into large mounds near mine openings, the dumping of mill tailings into rivers and lakes, and the excavation of open-pit mines have been the most common landscape alterations. Only recently have environmental historians and historical geographers such as Randall Rohe and Richard Francaviglia begun systematic studies of landscape evolution in mining districts.[34]

What is the likely future of the American copper industry in the twenty-first century and beyond? The only accurate prediction one can make is that any prediction will likely be off the mark. Barring the discovery of large new deposits or major breakthroughs in mining or processing technology (neither of which are very likely to occur), the U.S. copper industry will probably continue its long-term decline.

In a provocative study of the long-term future of the world copper industry, Robert B. Gordon and three co-authors predict dramatically rising costs of copper as a result of the exhaustion of the current "rich" copper deposits. The title of their study, *Toward a New Iron*

Age? Quantitative Modeling of Resource Exhaustion (1987) suggests the direction of their research. They combine the methodologies of materials modeling and linear programming to project the supply of copper (both newly mined and scrap), consumption, and prices to the year 2150. Their model allows for considerable substitution of other materials (especially aluminum) for copper, but assumes that technological advances in copper production would produce cost savings only large enough to counterbalance rising input costs.

Gordon and his coauthors predict that the rate of extraction of copper from the earth will peak around 2100 and then slowly decline. The copper ores of today, with an average of roughly 0.5 percent copper, will be exhausted by about 2070. Thereafter, copper will be extracted from common rock, with a maximum copper content of about 0.05 percent, driving up production costs dramatically.

Copper metal prices will rise sharply, perhaps as much as sixtyfold by 2070, producing intensive new efforts to recycle copper and find substitute materials. They predict that 90 percent of the products that use copper today will use substitute materials in the future. Copper will still be used in some electrical applications but only where its unique qualities justify the extraordinary cost. Gordon and his colleagues admit that their predictions would be altered significantly if economic growth is slower than their model assumes or if technological change is faster. With luck, perhaps both conditions will hold.[35]

Notes

Abbreviations Used in the Notes

ACM Co. Papers	Anaconda Copper Mining Company Papers, Montana Historical Society
C & H	Calumet and Hecla Mining Company
EMJ	*Engineering and Mining Journal*
PLMG	*Portage Lake Mining Gazette*
QMC records	Quincy Mining Company records, Copper Country Historical Collections, Michigan Technological University
Transactions, A.I.M.E.	*Transactions of the American Institute of Mining Engineers*

Introduction

1. T. A. Rickard, *A History of American Mining* (New York: McGraw Hill, 1932). General histories of western metal mining include Rodman Wilson Paul, *Mining Frontiers of the Far West, 1848–1880* (New York: Holt, Rinehart and Winston, 1963); Clarke C. Spence, *British Investments and the American Mining Frontier, 1860–1901* (Ithaca, N.Y.: Cornell University Press, 1958); and Otis C. Young Jr., *Western Mining* (Norman, Okla.: University of Oklahoma Press, 1970). Recent general works that focus on labor history include Ronald C. Brown, *Hard-Rock Miners: The Intermountain West, 1860–1920* (College Station: Texas A & M University Press, 1979); Richard E. Lingenfelter, *The Hardrock Miners: A History of the Mining Labor Movement in the American West, 1863–1893* (Berkeley: University of California Press, 1974); and Mark Wyman, *Hard Rock Epic: Western Miners and the Industrial Revolution, 1860–1910* (Berkeley: University of California Press, 1979). The classic study of labor in metal mining is Vernon H. Jensen, *Heritage of Conflict: Labor Relations in the Nonferrous Metals Industry up to 1930* (Ithaca, N.Y.: Cornell University Press, 1950).

2. The best histories of individual copper districts are William B. Gates Jr., *Michigan Copper and Boston Dollars: An Economic History of the Michigan Copper Mining Industry* (Cambridge, Mass.: Harvard University Press, 1951); Larry D. Lankton, *Cradle to Grave: Life, Work, and Death at the Lake Superior Copper Mines* (New York: Oxford University Press, 1991); and Michael P. Malone, *The Battle for Butte: Mining and Politics on the Northern Frontier, 1864–1906* (Seattle: University of Washington Press, 1981).

3. See Orris C. Herfindahl, *Copper Costs and Prices, 1870–1957* (Baltimore: John Hopkins University Press, 1959); Raymond F. Mikesell, *The World Copper Industry: Structure and Economic Analysis* (Baltimore: John Hopkins University Press, 1979);

Thomas R. Navin, *Copper Mining and Management* (Tucson: University of Arizona Press, 1978); and George H. Hildebrand and Garth L. Mangum, *Capital and Labor in American Copper, 1845–1990: Linkages between Product and Labor Markets* (Cambridge, Mass.: Harvard University Press, 1992).

Chapter 1. Foundations

1. James A. Mulholland, *A History of Metals in Colonial America* (University, Ala.: University of Alabama Press, 1981), 4–10, 18–25; and Edmund S. Morgan, "The Labor Problem at Jamestown, 1607–1618," *American Historical Review* 76 (June 1971): 597–600.

2. J. M. French, "The Simsbury Copper Mines," *New England Magazine*, old ser., 5 (March 1887): 430–33; Collamer M. Abbott, "Colonial Copper Mines," *William and Mary Quarterly* 27 (April 1970): 297–98; and Noah Amherst Phelps, *History of Simsbury, Granby and Canton from 1642 to 1845* (Hartford, Conn.: Case, Tiffany & Burnham, 1845), 117–18. For a detailed history of Newgate Prison, see Richard H. Phelps, *A History of Newgate of Connecticut, at Simsbury* (New York: J. Munsell, 1860). The same author produced a slightly different later version: *Newgate of Connecticut: Its Origin and Early History* (Hartford, Conn.: American Publishing Company, 1876).

3. J. H. Granbery, "The Schuyler Mine," *Journal of the Franklin Institute* 164 (July 1907): 13–15; and Elizabeth Marting, "Arent Schuyler and His Copper Mine," *Proceedings of the New Jersey Historical Society* (1947): 126–28.

4. Harry B. Weiss and Grace M. Weiss, *Old Copper Mines of New Jersey* (Trenton, N.J.: Past Time Press, 1963), 14–17.

5. Edmond Dale Daniel, "Robert Hunter Morris and the Rockey Hill Copper Mine," *New Jersey History* 92 (1974): 16–20; Weiss and Weiss, *Old Copper Mines of New Jersey*, 48–50, 78–79; and Nancy C. Pearre, "Mining for Copper and Related Minerals in Maryland," *Maryland Historical Magazine* 59 (March 1964): 20–21.

6. See French, "Simsbury Copper Mines," 430; Abbott, "Colonial Copper Mines," 297; Herbert P. Woodward, *Copper Mines and Mining in New Jersey*, Department of Conservation and Development, State of New Jersey, Bulletin 57, Geological Series (Trenton, N.J., 1944), 46–49; Daniel, "Robert Hunter Morris and the Rockey Hill Copper Mine," 16–20; and also Weiss and Weiss, *Old Copper Mines of New Jersey*, 79; and Pearre, "Mining for Copper and Related Minerals in Maryland," 20.

7. David J. Krause, *The Making of a Mining District: Keewenaw Native Copper, 1500–1870* (Detroit: Wayne State University Press, 1992), 39–40; *Dictionary of Canadian Biography* (Toronto: University of Toronto Press, 1966), 1:130–33, 590–92; Alexander Henry, *Travels and Adventures in Canada and the Indian Territories between the Years 1760 and 1776*, ed. James Bain (1809; reprint, Toronto: George Morang, 1901), 225–26; John Bartlow Martin, *Call It North Country: The Story of Upper Michigan* (New York: Knopf, 1944), 38; Lewis Beeson and Victor E. Lemmer, *The Effects of the Civil War on Mining in Michigan* (Lansing: Michigan Historical Commission, 1966), 2–5; and Robert James Hybels, "The Lake Superior Copper Fever, 1841–47," *Michigan History* 34 (June 1950): 100–101.

8. Joan Day, *Bristol Brass: A History of the Industry* (North Pomfret, Vt.: David &

Charles, 1973), 44–45; Henry Hamilton, *The English Brass and Copper Industries to 1800* (London: Longmans, Green & Company, 1926), 105, 144; Elizabeth B. Schumpeter, *English Overseas Trade Statistics, 1697–1808* (Oxford: Clarendon Press, 1960), 53–54, 63 (the copper equivalent of brass is assumed to be two-thirds of the weight of brass); and U.S. Bureau of the Census, *Historical Statistics of the United States: Colonial Times to 1970*, pt. 2 (Washington, D.C.: USGPO, 1975), 1184.

9. *Historical Statistics of the United States*, pt. 2, p. 1168; and Schumpeter, *English Overseas Trade Statistics, 1697–1808*, 63.

10. Henry J. Kauffman, *American Copper and Brass* (Camden, N.J.: Thomas Nelson & Sons, 1968), 63–142, 261–73.

11. Ibid., 143–250, 274–81; and James Leander Bishop, *A History of American Manufactures from 1608 to 1860*, 3d ed. (Philadelphia: Edward Young & Company, 1868), 1:574.

12. Denys B. Barton, *A History of Copper Mining in Cornwall and Devon* (Truro: Truro Bookshop, 1961), 91; and Charles Lemon, "The Statistics of the Copper Mines of Cornwall," *Journal of the Statistical Society of London* 1 (1838), reprinted in Roger Burt, ed., *Cornish Mining: Essays on the Organization of Cornish Mines and the Cornish Mining Economy* (Newton Abbot: David & Charles, 1969), 80–81. Barton's tables of output statistics refer to metallic copper but are erroneously labeled as statistics of *ore* production.

13. Robert Hunt, *British Mining* (London: Crosby Lockwood & Co., 1884), 892.

14. F. W. Gibbs, "Extraction and Production of Metals: Non-Ferrous Metals," in Charles Singer et al., *A History of Technology*, vol. 4: *The Industrial Revolution, c. 1750 to c. 1850* (New York: Oxford University Press, 1958), 126–29; Robert R. Toomey, *Vivian and Sons, 1809–1924: A Study of the Firm in the Copper and Related Industries* (New York: Garland Publishing, 1985), 277–80; and George Hammersley, "The Effect of Technical Change in the British Copper Industry between the Sixteenth and the Eighteenth Centuries," *Journal of European Economic History* 20 (Spring 1991): 161–62.

15. Toomey, *Vivian and Sons*, 281–84.

16. Edmund Newell, "'Copperopolis': The Rise and Fall of the Copper Industry in the Swansea District, 1826–1931," *Business History* 32 (July 1990): 76–77.

17. John Miers, *Travels in Chile and La Plata, 1819–1825* (London, 1826), 2:337–40, 387; H. Foster Bain and Thomas T. Read, *Ores and Industry in South America* (New York: Harper & Brothers, 1934), 211–12; Leland R. Pederson, *The Mining Industry of the Norte Chico, Chile* (Evanston, Illinois: Northwestern University Department of Geography, 1966), 195–202; Joanne Fox Przeworski, *The Decline of the Copper Industry in Chile and the Entrance of North American Capital, 1870–1916* (New York: Arno Press, 1980), 93–95; James Fred Rippy, *British Investment in Latin America, 1822–1949* (Minneapolis: University of Minnesota Press, 1959), 24; and John Rowe, *Cornwall in the Age of the Industrial Revolution* (Liverpool: Liverpool University Press, 1953), 141, 145.

18. Weiss and Weiss, *Old Copper Mines of New Jersey*, 18–21.

19. Charles Emil Peterson, "Notes on Copper Roofing in America to 1802," Society of Architectural Historians, *Journal* 24 (December 1965): 315–18.

20. John R. Harris, "Copper and Shipping in the Eighteenth Century," *Economic History Review*, 2d ser., 19 (December 1966): 550–55, 558, 564–67; Otis E.

Young Jr., "Origins of the American Copper Industry," *Journal of the Early Republic* 3 (1983): 126n. 22; Elva Tooker, *Nathan Trotter, Philadelphia Merchant, 1787–1853* (Cambridge, Mass.: Harvard University Press, 1955), 69; and John H. Morrison, *History of the New York Ship Yards* (1909; reprint, Port Washington: Kennikat Press, 1970), 18.

21. Mauer Mauer (*sic*), "Coppered Bottoms for the United States Navy, 1794–1803," United State Naval Institute *Proceedings* 71 (June 1945): 693–96.

22. Esther Forbes, *Paul Revere and the World He Lived In* (Cambridge, Mass.: Riverside Press, 1942), 368–72, 378–80, 407–8; and Isaac F. Marcosson, *Copper Heritage: The Story of Revere Copper and Brass* (New York: Dodd, Mead & Company), 38–40. Alfred A. Cowles asserts that the rolls were made in Maidenhead ("Copper and Brass," in Chauncey M. Depew, ed., *1795–1895: One Hundred Years of American Commerce* [1895, reprinted New York: Greenwood Press, 1968], 1:333), while Forbes gives Maidstone as the place of manufacture (*Paul Revere*, 408).

23. Marcosson, *Copper Heritage*, 41–43, 46.

24. Clayton C. Hall, ed., *Baltimore: Its History and Its People* (New York: Lewis Historical Publishing Company, 1912), 1:524; R. Brent Keyser, "Copper," in *Maryland: Its Resources, Industries, and Institutions* (Baltimore: Board of World's Fair Managers, 1893), 115–17; Ralph J. Robinson, "Maryland's 200-Year-Old Copper Industry," *Baltimore* (July 1939): 25–27; Kauffman, *American Copper and Brass*, 27, 29; Marcosson, *Copper Heritage*, 64–66; and the *Jeffersonian*, 26 October 1934, n.p. The administration accounts of the estate of Levi Hollingsworth are in the Hollingsworth Papers, Maryland Historical Society Archives, Baltimore.

25. Maxwell Whiteman, *Copper for America: The Hendricks Family and a National Industry* (New Brunswick, N.J.: Rutgers University Press, 1971), 38–40, 48–50, 109–15; and Young, "Origins of the American Copper Industry," 131.

26. U.S. Congress, *New American State Papers: Manufactures* (Wilmington, Del.: Scholarly Resources, 1972), 1:63; and Bishop, *History of American Manufactures*, 2:54, 228, 242, 375, 614.

27. U.S. Congress, *Digest of Manufactures, As Ordered by Congress, March 30, 1822, Comprising a Statement of the Manufacturing Establishments of the United States, American State Papers: Finance* (Washington, D.C.: Gales & Seaton, 1858), 4:42, 100, 138–43. The Gunpowder Copper Works in Maryland does not appear at all, reflecting the haphazard nature of the returns.

28. Ibid., 154–57.

29. Whiteman, *Copper for America*, 160–61, 165–68.

30. Robinson, "Maryland's 200-Year-Old Copper Industry," 27; Hill, *Baltimore*, 1:524; James A. Douglas, "Historical Sketch of Copper Smelting in the United States," *Mineral Industry, 1895*, 271; and Tooker, *Nathan Trotter*, 246n. 28.

31. Commander William Tingey, Washington Navy Yard, to Messrs. J., W., and E. Patterson, Merchants, Baltimore, 14 March 1826, Patterson Papers, Maryland Historical Society Archives, Baltimore; Robinson, "Maryland's 200-Year-Old Copper Industry," 27; and Tooker, *Nathan Trotter*, 93. Tooker identified Ellicott's firm as the Avalon Company. Tingey's letter indicated that Ellicott was to supply iron and copper for two sloops of war.

32. For a more complete discussion of the failings of the U.S. trade statistics before 1821, see U.S. Bureau of the Census, *Historical Statistics of the United States:*

Colonial Times to 1970 (Washington, D.C.: USGPO, 1975), 2:876–77; and Douglass C. North, "The United States Balance of Payments, 1790–1860," in National Bureau of Economic Research, *Studies in Income and Wealth,* vol. 24: *Trends in the American Economy in the Nineteenth Century* (Princeton, N.J.: Princeton University Press, 1960), 590–94.

33. Kauffman, *American Copper and Brass,* 174–281.

34. Whiteman, *Copper for America,* 47; William G. Lathrop, *The Brass Industry in the United States,* rev. ed. (Mt. Carmel, Conn.: William G. Lathrop, 1926), 47; *New American State Papers: Commerce and Navigation* (Wilmington, Del.: Scholarly Resources, 1973), 7:312–13, 318–19; and Anne Bezanson, Robert D. Gray, and Miriam Hussey, *Wholesale Prices in Philadelphia, 1784–1861* (Philadelphia: University of Pennsylvania Press, 1937), 2:53–54. The official total for copper and brass imports for the year ending 30 September 1821 was $704,717, but the total included imports of tin bars, so it was adjusted downward. When tin bars were listed separately starting in 1825, the value of imported tin was between $50,000 and $60,000 per annum.

35. The foreign-trade figures in this segment come from *The New American State Papers: Commerce and Navigation,* vols. 7, 9, 10, 12, 15.

36. Lathrop, *Brass Industry in the United States,* 52; and Bezanson, *Wholesale Prices in Philadelphia, 1784–1861,* 2:53–54. Bezanson's prices are newspaper quotations, which she argues are quite reliable, based on comparisons with independent sources. For the case of copper sheathing, she compared the monthly price records of Nathan Trotter and Company, a prominent Philadelphia metal merchant, with the newspaper quotations and found the later to be reliable. For a detailed discussion and comparison of the price series, see Bezanson, *Wholesale Prices in Philadelphia, 1784–1861,* 1:333–35.

37. *New American State Papers: Commerce and Navigation,* vols. 7, 9, 10, 12, and 15.

38. Christopher J. Schmitz, *World Non-Ferrous Metal Production and Prices, 1700–1976* (London: Cass, 1979), 64; Arthur Gayer, W. W. Rostow, and Anna Schwartz, *The Growth and Fluctuations of the British Economy, 1790–1850,* 2 vols. (New York: Oxford University Press, 1953), Microfilm Supplement, 904; and Hunt, *British Mining,* 830.

39. Collamer M. Abbott has published a dozen articles on eastern copper, cited in this and other chapters. He has also produced a substantial unpublished MS, "Appalachian Copper: Red Metal Mining in the Eastern United States," which he graciously allowed me to see.

40. Collamer M. Abbott, "Boston Money and Appalachian Copper," *Michigan History* 55 (Fall 1971): 219–21, 227; "Isaac Tyson, Jr., Pioneer Industrialist," *Business History Review* 42 (Spring 1968): 71–72; "Early Copper Smelting in Vermont," *Vermont History* 33 (1965): 234–40; and "Green Mountain Copper: The Story of Vermont's Red Metal," unpublished MS, 1964, 7–19.

41. Child, *Gazetteer of Orange County, Vermont,* 15–17: Collamer M. Abbott, "Thomas Pollard, Cornish Miner," *Rev. Int. d'Hist. de la Banque* (Italy), 6 (1973): 171; and Walter Harvey Weed, *Notes on the Copper Mines of Vermont,* U.S. Geological Survey, Bulletin No. 225 (Washington, D.C.: USGPO, 1904), 194–95.

42. Epaphroditus Peck, *A History of Bristol, Connecticut* (Hartford: Lewis Street Bookshop, 1932), 135; Codman Hislop, *Eliphalet Nott* (Middletown, Conn.: Wes-

leyan University Press, 1971), 435, 520, 522–23; and Milo L. Norton, "Copper Mines in Bristol," in *Bristol, Connecticut, or the "New Cambridge"* (Hartford: City Printing Company, 1907), 441.

43. Benjamin F. Silliman Jr. and Josiah Dwight Whitney, *Report of an Examination of the Bristol Copper Mine in Bristol, Conn., August 1855* (New Haven: T. J. Stafford, 1855), 12, 21; Silliman and Whitney, "Notice of the Geological Position and Character of the Copper Mine of Bristol, Connecticut," *American Journal of Science and the Arts,* ser. 2, 20 (November 1855): 366–68; Gerald T. White, *Scientists in Conflict: The Beginnings of the Oil Industry in California* (San Marino, Calif.: the Huntington Library, 1968), 43; Charles R. Harte, "Connecticut's Iron and Copper," Connecticut Society of Civil Engineers, *Annual Report* (New Haven: Quinnipiak Press, 1944), 150–52; Peck, *History of Bristol,* 136–37; and Hislop, *Eliphalet Nott,* 530–31, 534. Hislop identified the "J. D. Whitney" who was involved at the Bristol mine as James Dana Whitney, but he was in fact Josiah Dwight Whitney.

44. Abbott, "Isaac Tyson, Jr.: Pioneer Industrialist," 71–74. Many of the deeds, agreements, and contracts relating to Tyson's chromium and copper mines in Maryland are found in the Tyson Record Book, 1826–49, in the Maryland Historical Society Archives in Baltimore.

45. Pearre, "Mining For Copper and Related Minerals in Maryland," 22–28; Robinson, "Maryland's 200-Year-Old Copper Industry," 25; and Whitney, *Metallic Wealth of the United States,* 317–19.

46. Weiss and Weiss, *Old Copper Mines of New Jersey,* 22–23, 30–37, 54–61.

47. Whitney, *Metallic Wealth of the United States,* 328.

48. Clarence S. Ross, *Origin of the Copper Deposits of the Ducktown Type in the Southern Appalachian Region,* U.S. Geological Survey Professional Paper 179 (Washington, D.C.: USGPO, 1935), 94–95; Robert E. Barclay, *Ducktown Back in Raht's Time* (Chapel Hill: University of North Carolina Press, 1946), 44–45; Carl Henrich, "The Ducktown Ore-Deposits and the Treatment of the Ducktown Copper-Ores," *Transactions, A.I.M.E.* 25 (1896): 177–79; Walter Harvey Weed, *Copper Deposits of the Appalachian States,* U.S. Geological Survey, Bulletin 455 (Washington, D.C.: USGPO, 1911), 152; and S. W. McCallie, "The Ducktown Copper Mining District," *EMJ* 74 (4 October 1902): 439.

49. "Copper Ore and Cotton: Dangerous Freight," *Hunt's Merchant Magazine and Commercial Review* 33 (September 1855): 394.

50. Barclay, *Ducktown Back in Raht's Time,* 77–84. The Burra Copper Company took its name from a highly successful Australian mine of the same name, launched in 1845. There was no other connection between the two ventures.

51. Ibid., 87–95; McCallie, "The Ducktown Copper Mining District," 439; and Henrich, "The Ducktown Ore-Deposits," 181.

52. Lathrop, *Brass Industry in the United States,* 42, 43, 54–56; and Robert Glass Cleland, *A History of Phelps Dodge, 1834–1950* (New York: Alfred A. Knopf, 1952), 4–12, 17, 23–26.

53. Lathrop, *Brass Industry in the United States,* 56–64, 100.

54. James A. Douglas, "Historical Sketch of Copper Smelting in the United States," *Mineral Industry, 1895,* 275; and Young, "Origins of the American Copper Industry," 130–31.

55. "Smelting Copper Ore," *American Journal of Science and the Arts* 54 (1847):

292; and Thomas Egleston, "The Point Shirley Copper Works," *School of Mines Quarterly* 7 (July 1886): 360–61.

56. "The Baltimore Copper Works," *EMJ* 32 (6 August 1881): 87; Douglas, "Historical Sketch of Copper Smelting in the United States," 272–73; and Robinson, "Maryland's 200-Year-Old Copper Industry," 27–28.

57. Robinson, "Maryland's 200-Year-Old Copper Industry," 27–28; and David S. Van Tassel, "The Baltimore Refinery of the American Smelting and Refining Company: A History," unpublished report prepared for ASARCO, Inc. (1976), 28.

58. Young, "Origins of the American Copper Industry," 131; *American Mining Index*, 6 June 1866; and Douglas, "Historical Sketch of Copper Smelting in the United States," 274–75.

59. Donald Chaput, *The Cliff: America's First Great Copper Mine* (Kalamazoo, Mich.: Sequoia Press, 1971), 22, 25, 52, 58.

60. James B. Cooper, "Historical Sketch of Smelting and Refining Lake Copper," *Proceedings of the Lake Superior Mining Institute* 7 (1901): 44–45.

61. George H. Thurston, *History of Allegheny County, Pennsylvania* (Chicago: A. Warner & Co., 1889), 255; Benjamin S. Johns, "Henry Johns, an Early American Copper Refiner," *EMJ* 93 (15 June 1912): 1183; "Curtis G. Hussey," *Magazine of Western History* 3 (February 1886): 338–40; Egleston, "Copper Refining in the United States," 680–89; and Chaput, *Cliff*, 65–66.

62. Cooper, "Historical Sketch of Smelting and Refining Lake Copper," 45; "J. G. Hussey & Co.'s Copper Smelting Works," *Lake Superior Miner*, 31 July 1858, 371; and Thomas Egleston, "Copper Refining in the United States," 681–88.

63. Silas Farmer, *History of Detroit and Wayne County and Early Michigan* (Detroit: Silas Farmer & Company, 1890), 816–17; Lathrop, *Brass Industry in the United States*, 79, 85; "Detroit and Waterbury Copper Smelting Works," *Mining Magazine* 1 (1853): 298; and "The Detroit & Lake Superior Copper Co.'s Smelting Works," *Collections and Researches Made by the Michigan Pioneer and Historical Society*, 28 (Lansing: Robert Smith Printing Company, 1900), 648, 651.

64. Cooper, "Historical Sketch of Smelting and Refining Lake Copper," 46; "The Detroit and Lake Superior Copper Co.'s Smelting Works," 650–51; and *Manufactures of the United States in 1860; Compiled From the Original Returns of the Eighth Census* (Washington, D.C.: USGPO, 1865), 43, 67, 221, 245, 272, 337, 447, 494, 598, 609.

65. Toomey, *Vivian and Sons*, 289; *New American State Papers: Commerce and Navigation*, 29:176–77; 32:520–21; 39:175, 177; 44:177, 185.

66. *New American State Papers: Commerce and Navigation*, vols. 39, 44.

67. Hoval A. Smith, *American Copper Production and History of a Copper Tariff and Other Copper Tariff Details* (Miami, Ariz.: Arizona Silver Belt, 1932), 7–14.

Chapter 2. Michigan Copper through the Civil War

1. George A. West, *Copper: Its Mining and Use by the Aborigines of the Lake Superior District; Report of the McDonald-Massee Isle Royale Expedition, 1928*, originally published in the *Bulletin of the Public Museum of the City of Milwaukee*, vol. 10, 19 May 1929 (reprint, Westport, Conn.: Greenwood Press, 1970), 45–47; John R.

Halsey, "Miskwabik—Red Metal: Lake Superior Copper and the Indians of Eastern North America," *Michigan History* 67 (September/October 1983): 33, 37, 39; Ronald J. Mason, *Great Lakes Archaeology* (New York: Academic Press, 1981), 181–98, 373–80, 400–405; James B. Griffin, *Lake Superior Copper and the Indians: Miscellaneous Studies in Great Lakes Prehistory* (Ann Arbor: University of Michigan Press, 1961), 133; and R. L. Rickard, "Pre-Columbian Copper Mining in North America," in the Smithsonian Institution, *Annual Report for 1892* (Washington, D.C.: Smithsonian Institution, 1893), 179–80.

2. Charles Whittlesey, "Ancient Mining on the Shores of Lake Superior," *Smithsonian Contributions to Knowledge*, vol. 13, contribution no. 155 (Washington, D.C.: Smithsonian Institution, 1863), 6–22.

3. Halsey, "Miskwabik—Red Metal," 39.

4. Richard G. Bremer, "Henry Rowe Schoolcraft: Explorer in the Mississippi Valley, 1818–1832," *Wisconsin Magazine of History* 66 (Autumn 1982): 45–47; Henry R. Schoolcraft, "Account of the Native Copper on the Southern Shore of Lake Superior, with Historical Citations and Miscellaneous Remarks, In a Report to the Department of War," *American Journal of Science and the Arts* 3 (1821): 201–16; and Schoolcraft, *Narrative Journal of Travels Through the Northwestern Regions of the United States Extending from Detroit through the Great Chain of American Lakes to the Sources of the Mississippi River in the Year 1820*, edited by Mentor L. Williams (East Lansing: Michigan State University Press, 1953), 22–23.

5. "Proceedings of Detroit Meeting: Memorial to Congress by Citizens of the Territory," 28 October 1831, in Clarence E. Carter, ed., *Territorial Papers of the United States: The Territory of Michigan*, vol. 12: *1829–1837* (Washington, D.C.: USGPO, 1945), 367; Philip P. Mason, ed., *Schoolcraft's Expedition to Lake Itasca: The Discovery of the Source of the Mississippi* (East Lansing, Mich.: Michigan State University Press, 1958), 179–81, 230–31, 282–84, 290–94, 366; and David J. Krause, *The Making of a Mining District: Keweenaw Native Copper, 1500–1870* (Detroit: Wayne State University Press, 1992), 103–11.

6. Krause, *The Making of a Mining District*, 90–93, 113–17; and Edsel K. Rintala, *Douglass Houghton: Michigan's Pioneer Geologist* (Detroit: Wayne State University Press, 1954), 32–58.

7. Krause, *The Making of a Mining District*, 130–34; *Reports of the First, Second, and Third Meetings of the Association of American Geologists and Naturalists* (Boston: Gould, Kendall, & Lincoln, 1843), 35–38; "Proceedings of the Association of American Geologists and Naturalists," *American Journal of Science and the Arts* 41 (1841): 183–86; Douglass Houghton to William Woodbridge, 26 December 1840, in George N. Fuller, ed., *Geological Reports of Douglass Houghton* (Lansing: Michigan Historical Commission, 1928), 476–77; Douglass Houghton to Augustus Porter, 26 December 1840, in Alvah Bradish, *Memoir of Douglass Houghton, First State Geologist of Michigan* (Detroit: Raynor and Taylor, 1889), 113–16; and *Detroit Daily Advertiser*, 20 May 1841.

8. Lewis Beeson and Victor F. Lemmer, *The Effects of the Civil War on Mining in Michigan* (Lansing: Michigan Historical Commission, 1966), 8–9; Krause, *The Making of a Mining District*, 136–38; and Robert James Hybels, "The Lake Superior Copper Fever, 1841–1847," *Michigan History* 34 (1950): 224.

9. An excellent summary of federal law and policy for mineral lands is Rob-

ert W. Swenson, "Legal Aspects of Mineral Resources Exploitation," in Paul W. Gates, *History of Public Land Law Development* (Washington, D.C.: USGPO, 1968), 699–764. More detailed studies of leasing include John Wills Taylor, "Reservation and Leasing of the Salines, Lead, and Copper Mines of the Public Domain" (Ph.D. thesis, University of Chicago, 1930); and James E. Wright, *The Galena Lead District: Federal Policy and Practice, 1824–1847* (Madison: University of Wisconsin Press, 1966). For details of the early land speculation, see Lawrence Fadner, comp., *Fort Wilkins, 1843, and the U.S. Mineral Land Agency, Copper Harbor, Michigan, Lake Superior* (New York: Vantage Press, 1966); James Fisher, "Historical Sketch of the Lake Superior Copper District," *Proceedings of the Lake Superior Mining Institute* 27 (1929): 62, 64; and Hybels, "Lake Superior Copper Fever," 111–13, 237–41.

10. U.S., 28th Cong., 2d sess., Senate Exec. Doc. 175 (1845), 1–22; Fadner, *Fort Wilkins*, 155–254; W. L. Marcy, *Report on the Mineral Regions of Lake Superior, February 23, 1846*, 29th Cong., 1st Session, Sen. Doc. 160, 4:19; and James H. Rolfe, *Report on the Sale of Mineral Lands to the Committee on Public Lands, May 4, 1846*, 29th Cong., 1st sess., House Report 591, 3:45.

11. Hybels, "Lake Superior Copper Fever," 234–43, 317–18.

12. *Niles'National Register* 69 (May 1845): 192. Similarly, a Cornish miner's report of large silver deposits near Eagle River in January 1846, was reprinted without comment in the *American Railroad Journal* 19 (1846): 695.

13. Hybels, "Lake Superior Copper Fever," 239, 241, 311; *Mineral Statistics for 1880*, 152; John R. St. John, *A True Description of the Lake Superior Country* (New York: William H. Graham, 1846), 78–91; Jacob Houghton Jr. and T. W. Bristol, *Reports on the Mineral Region of Lake Superior With a Correct Map of the Same and a Chart of Lake Superior* (1846), 92–109; *Lake Superior News and Miners' Journal*, 11 July, 24 October 1846; and Charles Lanman, *A Summer in the Wilderness, Embracing a Canoe Voyage up the Mississippi and around Lake Superior* (New York: Appleton, 1847), 154–55.

14. Beeson and Lemmer, *Effects of the Civil War*, 13; Hybels, "Lake Superior Copper Fever," 13; and Swenson, "Legal Aspects," 706–7.

15. John R. St. John, *A True Description of the Lake Superior Country; Its Rivers, Coasts, Bays, Harbours, Islands, and Commerce. With Bayfield's Chart; Also a Minute Account of the Copper Mines and Working Companies, Accompanied by a Map of the Mineral Regions, Showing, by their Number and Place, All the Different Mineral Locations: and Containing A Concise Mode of Assaying, Treating, Smelting, and Refining Copper Ores* (New York: William H. Graham, 1846); and Jacob Houghton Jr. and T. W. Bristol, *Reports on the Mineral Region of Lake Superior With a Correct Map of the Same and a Chart of Lake Superior* (1846).

16. Mentor L. Williams, "Horace Greeley and Michigan Copper," *Michigan History* 34 (June 1950): 120–32. The *Lake Superior News* moved to Sault Ste. Marie in 1847 and remained there until November 1855, when it began publishing in Marquette.

17. William B. Gates Jr., *Michigan Copper and Boston Dollars: An Economic History of the Michigan Copper Mining Industry* (Cambridge, Mass.: Harvard University Press, 1951), 197; and *Mineral Statistics for 1880*, 21; and J.D.B. DeBow, *Statistical View of the United States: Being a Compendium of the Seventh Census* (Washington, D.C.: A.O.P. Nicholson, Public Printer, 1854), 254. For a more detailed discussion

of the investment patterns, see Charles K. Hyde, "From 'Subterranean Lotteries' to Orderly Investment: Michigan Copper and Eastern Dollars, 1841–1865," *Mid-America* 66 (January 1984): 3–16.

18. *Mineral Statistics for 1880,* insert fol. 152; and Gates, *Michigan Copper,* 197.

19. Donald Chaput, *The Cliff: America's First Copper Mine* (Kalamazoo, Mich.: Sequoia Press, 1971), 18–25.

20. Ibid., 36–37; *Mineral Statistics for 1880,* 21; and Gates, *Michigan Copper,* 216–17.

21. Gates, *Michigan Copper,* 216–17; Angus Murdoch, *Boom Copper: The Story of the First U.S. Mining Boom* (New York: Macmillan, 1943), 38, 97; and *Mineral Statistics for 1880,* 68–69, 74–75, and insert fol. 152.

22. *Mineral Statistics for 1880,* 36–41, 44–47, 152.

23. Gates, *Michigan Copper,* 11.

24. Jackson, *Report on the Geological and Mineralogical Survey* (1849); and John W. Foster and Josiah D. Whitney, *Report on the Geology and Topography of a Portion of the Lake Superior Land District in the State of Michigan,* pt. 1, *Copper Lands,* 31st Cong., 1st sess., vol. 9, no. 69 (Washington, D.C.: House of Representatives, 1850); Robert E. Clarke, "Notes from the Copper Region," *Harper's New Monthly Magazine* 6 (March and April 1853): 438–48, 577–88; and Josiah D. Whitney, *The Metallic Wealth of the United States, Described and Compared with That of Other Countries* (Philadelphia: Lippencott, 1854). The complete title was *Mining Magazine and Journal of Geology, Mineralogy, Metallurgy, Chemistry, and the Arts.* In addition, an extensive description of the Lake Superior mines by the Frenchman Louis Edouard Rivot, translated as "Visit to the Lake Superior Region, 1854," appeared in the *Mining Magazine* 6 (1855): 27–28, 99–106, 207–13, 414–18; 7 (1856): 249–55, 359–67; 8 (1857): 60–65.

25. Gates, *Michigan Copper,* 32–33.

26. Ibid., 34–35.

27. Larry D. Lankton and Charles K. Hyde, *Old Reliable: An Illustrated History of the Quincy Mining Company* (Franklin, Mich.: Four Corners Press, 1982), 5–11, 15–17.

28. John Wills Taylor, "Reservation and Leasing of the Salines, Lead, and Copper Mines of the Public Domain," 267; Murdoch, *Boom Copper,* 48–49; and chap. 13; Hybels, "Lake Superior Copper Fever," 241–42; Gates, *Michigan Copper,* 32, 239n. 123; and Abbott, "Boston Money and Appalachian Copper," 217–18. Gates, whose book was based on his Ph.D. thesis in economics at the University of Chicago in the late 1940s, was unaware of Hybels, his contemporary working in Chicago's history department.

29. Houghton and Bristol, *Reports on the Mineral Region of Lake Superior,* 92–109; and *Lake Superior News and Miners' Journal* 11 July 1846.

30. *Lake Superior Miner* 27 October 1855; and *Mineral Statistics for 1880,* insert fol. 152.

31. Joseph G. Martin, *Martin's Boston Stock Market: Eighty-Eight Years, From January 1798 to January, 1886* (Boston: the author, 1886), 136; and R. G. Dun & Company Collections, Baker Library, Harvard University School of Business Administration, *Michigan,* vol. 24, pp. 261–74.

32. Gates argues that the cost data were rarely of much value to stockholders; *Michigan Copper,* 22. This was true for the extremely sketchy Calumet and Hecla Mining Company annual reports but was generally not the case for those issued

by other companies. The most extensive collections of Michigan mining company reports are found in the Baker Library at Harvard University and in the Copper Country Historical Collections at Michigan Technological University, Houghton, Michigan. LeRoy Barnett, *Mining in Michigan: A Catalogue of Company Publications* (Marquette: Northern Michigan University Press, 1983), is a valuable aid in locating these company reports.

33. *Mineral Statistics for 1880,* insert fol. 152.

34. Gates, *Michigan Copper,* 15, 216; Martin, *Martin's Boston Stock Market,* 136, 145; Barnett, *Mining in Michigan,* passim; and *PLMG,* 28 January 1865. The quotation is from the *PLMG,* 28 January 1865.

35. Gates, *Michigan Copper,* 16–17, 197; and *Mineral Statistics for 1880,* insert fol. 152.

36. *Mineral Statistics for 1880,* insert fol. 152; Barnett, *Mining in Michigan,* passim; *Hunt's Merchants' Magazine and Commercial Review* 54 (March 1866): 220; and *PLMG,* 9 September 1865.

37. C. Harry Benedict, in *Red Metal: The Calumet and Hecla Story* (Ann Arbor: University of Michigan Press, 1951), 6–10, cites the *Portage Lake Mining Gazette* of 23 December 1865 but fails to note that the figures for the amount of capital paid in was adjusted downward by $708,756 in the *Portage Lake Mining Gazette* of 13 January 1866. The amount reinvested is a crude estimate at best. Seven of the larger producers (Cliff, Copper Falls, Franklin, Minesota, Northwestern, Pewabic, and Quincy) "plowed back" a total of $2.4 million of their sales receipts.

38. *Hunt's Merchants' Magazine and Commercial Review* 54 (March 1866): 220 gives the data on capital and dividends. The information on output and revenues are from *Mineral Statistics for 1880,* inset fol. 152.

39. Thomas Egleston, "Copper Mining on Lake Superior," *Transactions, A.I.M.E.* 6 (1879): 285–88; and Lankton and Hyde, *Old Reliable,* 8, 154.

40. Victor Lemmer and Lewis Beeson, *Effects of the Civil War on Mining in Michigan* (Lansing: Michigan Historical Commission, 1966), 16–17; Butler and Burbank, *Copper Deposits of Michigan,* 65.

41. Chaput, *Cliff,* 34–44, 52; *Ninth Report to the Stockholders of the Franklin Mining Company, For the Year Ending December 31, 1868* (Boston: Alfred Mudge & Son, Printers, 1869), 18–19; and *Report of the Directors to the Stockholders of the Pewabic Mining Company, For the Year Ending December 31, 1869* (Boston: Alfred Mudge & Son, Printers, 1869), 16–17.

42. William Pryce, *Mineralogia Cornubiensis: A Treatise on Minerals, Mines, and Mining* (London: James Phillips, 1778), 137–72; John R. Leifchild, *Cornwall: Its Mines and Miners* (London: Longman, Brown, Green, Longmans, and Roberts, 1857), 136–61; and A. K. Hamilton Jenkin, *The Cornish Miner: An Account of His Life Above and Underground from Early Times* (London: George Allen & Unwin, 1927), 92–104. Transplanted Cornish mining practices included the use of the fathom as the unit of measurement underground and the practice of calling the chief underground manager the mining captain. Terms used in Michigan copper mines, including *adit, winze, stope, raise, whim, kibble,* and *stamp,* are found in a glossary in Pryce, *Mineralogia Cornubiensis* (1778), 315–31.

43. For a more detailed description of mining technology, see Lankton and Hyde, *Old Reliable,* 22–27.

44. Ibid., 27–32.

45. Gates, *Michigan Copper*, 24–30.

46. The best study of labor organization and conditions in Cornish mines is John Gardner Rule, "The Labouring Miners in Cornwall, c. 1740–1870: A Study in Social History" (Ph.D. thesis, University of Warwick, 1971), 9, 27–32. According to Rule, a compilation prepared in 1787 showed a workforce of 7,196, with 37 per cent women and children. Similarly, Charles Lemon shows the copper mines as employing more than 27,000 workers in 1836, with 41 per cent of them women and children; "The Statistics of the Copper Mines of Cornwall," *Journal of the Statistical Society of London* 1 (1838) reprinted in Burt, ed., *Cornish Mining*, 70–75.

47. U.S. Bureau of the Census, *Ninth Census, 1870*, vol. 3: *The Statistics of the Wealth and Industry of the United States* (Washington: USGPO, 1872), 767; and Raphael Pumpelly, *Report on the Mining Industries of the United States*, U.S. Bureau of the Census, *Tenth Census, 1880*, vol. 15, pt. 2 (Washington, D.C.: USGPO, 1886), 798.

48. Pryce, *Mineralogia Cornubiensis* (1778), 180–85, 188–90; John Taylor, "On the Economy of the Mines of Cornwall and Devon," *Transactions of the Geological Society*, 12 (1814): 309–27; Leifchild, *Cornwall: Its Mine and Miners* (1857), 143–48; and Jenkin, *Cornish Miner* (1927), 134–40, 204–10, and 224–32, essentially describe the same systems for organizing work.

49. Lankton and Hyde, *Old Reliable*, 10, 16; *Lake Superior Miner*, 18 July 1857; and Chaput, *Cliff*, 100–101.

50. In Cornwall, tributing "pares" often had only two men, because they would work only one shift. But tutwork pares usually had at least six men and often as many as twelve; see Rule, "The Labouring Miner," 38. Quincy and other mining companies kept detailed records of each contract, including the names of all the contractors, their gross earnings, expenses, net earnings, and the division of earnings.

51. Alfred Nicholls suggests that the mine captain simply announced the rates; "Saarvey Day," in Roy W. Drier, ed., *More Copper Country Tales* (Calumet, Mich.: Roy Drier, 1968), 11–32. The work of the mining captain is described in Thomas Egleston, "Copper Mining on Lake Superior," *Transactions, A.I.M.E.* 6 (1877–78): 278–81. The contract auction, sometimes called a "Dutch auction," in Cornwall is described in detail by Langford Lovell Price, *"West Barbary"; or, Notes on the System of Work and Wages in the Cornish Mines* (London: H. Frowde, 1891), reprinted in Burt, ed., *Cornish Mining*, 134–35.

52. Egleston, "Copper Mining on Lake Superior," 280. The traditional Cornish system of awarding tutwork contracts by auction had disappeared long before Egleston wrote in the late 1870s. A description of the labor system in use a quarter century earlier suggests that the auction system was never used in Michigan; see "Journal of Copper Mining Operations," *Mining Magazine* 1 (September 1853): 294. For details of the dispute at the Northwest Mining Company, see Franklin Hopkins to Samuel M. Day, 7 May 1855, Northwest Mining Company Letterbook, 1851–61, Bentley Historical Library, University of Michigan.

53. QMC records, Time Book, 1851–55.

54. QMC records, Returns of Labor, 1857–64, and *Annual Reports*. For a detailed discussion of the occupations at the Quincy Mine, including the distribution of workers and their pay rates, see Charles K. Hyde, "An Economic and

Business History of the Quincy Mining Company," unpublished report for the Historic American Engineering Record, National Park Service (HAER No. MI-2).

55. QMC records, Payroll Accounts, June 1865. A detailed listing is available from the author upon request.

Chapter 3. Michigan Copper in Prosperity and Decline

1. Gates, *Michigan Copper*, 198–99.

2. Ibid., 39–40, 203, 216–17; and Lankton and Hyde, *Old Reliable*, 18.

3. Gates, *Michigan Copper*, 40, 197, 230.

4. C. Harry Benedict, *Red Metal: The Calumet and Hecla Story* (Ann Arbor: University of Michigan Press, 1952), 29, 33, 42–43, 47–55, 63–65. This undocumented and uncritical volume is the only available history of this important producer. Benedict was a lifelong employee of the Calumet and Hecla Company, which underwrote the entire cost of publication of the book. For Hulbert's account of his work, see *"Calumet-Conglomerate," An Exploration and Discovery Made by Edwin J. Hulbert, 1854 to 1864* (Ontonagon, Mich.: Ontonagon-Miner Press, 1893), 27–29, 49, 105–20.

5. Edward B. Smith & Company, *Calumet and Hecla Mining Company* (New York, 1923), 18, 20; Benedict, *Red Metal*, 40–41, 56–57, 69–75; and Gates, *Michigan Copper*, 44.

6. Benedict, *Red Metal*, 80; and *Annual Report of the Commissioner of Mineral Statistics of the State of Michigan for 1880*, 121, 131.

7. Benedict, *Red Metal*, 75; Lankton and Hyde, *Old Reliable*, 152; and *The Mineral Industry, 1892*, 139, 230.

8. Rothwell, *Mineral Industry, 1892*, 196–97; Gates, *Michigan Copper*, 58; and Lankton and Hyde, *Old Reliable*, 61.

9. Gates, *Michigan Copper*, 84; and Lankton and Hyde, *Old Reliable*, 66–68.

10. Larry D. Lankton, "The Machine under the Garden: Rock Drills Arrive at the Lake Superior Copper Mines, 1868–1883," *Technology and Culture* 24 (January 1983): 9–22.

11. QMC records, *Annual Report for 1877*, 4; *Annual Report for 1882*, 3; and Lankton, "The Machine under the Garden," 22.

12. Lankton, "The Machine under the Garden," 27–28; and QMC records, Cost Sheets, 1893–1900.

13. *Sixth Annual Report of the Bureau of Labor and Industrial Statistics, February 1, 1889* (Lansing, Mich.: Thorp & Godfrey, 1889), 94–220; and U.S. Bureau of the Census, *Mines and Quarries, 1902* (Washington, D.C.: USGPO, 1905), 474.

14. Gates, *Michigan Copper*, 55–56; *Annual Report of the Commissioner of Mineral Statistics of the State of Michigan for 1880*, 218; and *State of Michigan, Mines and Mineral Statistics for 1891* (Lansing: Robert Smith & Company, 1892), 41.

15. Gates, *Michigan Copper*, 28–29, 42–43, 72; Egleston, "Copper Refining in the United States," 681; and Cooper, "Historical Sketch of Smelting and Refining Lake Copper," 46.

16. Egleston, "Copper Refining in the United States," 685; Raphael Pumpelly, *Report on the Mining Industries of the United States*, U.S. Bureau of the Census, *Tenth Census, 1880*, vol. 15, pt. 3 (Washington, D.C.: USGPO, 1886), 798–800; Lankton

and Hyde, *Old Reliable,* 90; and Cooper, "Historical Sketch of Smelting and Refining Lake Copper," 46.

17. Donald Chaput, *Hubbell: A Copper Country Village* (Lake Linden, Mich.: John H. Foster Press, 1986), 16–19; and Benedict, *Red Metal,* 96–99.

18. Alfred B. Lindley, "The Copper Tariff of 1869," *Michigan History* 35 (March 1951): 2–3, 9–10; and U.S. Treasury Department, United States Revenue Commission, *Special Report No. 11: Copper Mining and Manufacture* (Washington, D.C.: USGPO, 1866), 17–21, 29–32.

19. Gates, *Michigan Copper,* 216–17; and Hoval A. Smith, *American Copper Production and History of a Copper Tariff and Other Copper Tariff Details* (Miami, Ariz.: Arizona Copper Tariff Commission, 1932), 80.

20. *American Mining Index,* 6 June 1867; "The Baltimore Copper Works," *EMJ* 32 (6 August 1881): 87–88; R. Brent Keyser, "Copper," in *Maryland: Its Resources, Industries and Institutions* (Baltimore: Board of World's Fair Managers of Maryland, 1893), 116–18; and T. Egleston, "The Point Shirley Copper Works," *School of Mines Quarterly* 7 (July 1886): 361.

21. J. Ross Browne, *Report on the Mineral Resources of the States and Territories West of the Rocky Mountains* (Washington, D.C.: USGPO, 1868), 207–19; Gates, *Michigan Copper,* 197; and U.S. Bureau of the Census, *Ninth Census, 1870,* vol. 3: *The Statistics of the Wealth and Industry of the United States* (Washington, D.C.: USGPO, 1872), 767.

22. Browne, *Mineral Resources of the States and Territories West of the Rocky Mountains,* 211–18; and *Commerce and Navigation of the United States for the Year Ending June 30, 1869,* 41st Cong., 2d sess., House Exec. Docs., vol. 15, Serial 1429 (Washington: USGPO, 1870), 378.

23. *Mineral Industry, 1892,* 129; *PLMG,* 2 June 1870; and Gates, *Michigan Copper,* 197.

24. QMC records, New York Office Records, untitled documents dated "New York, Tuesday March 1, 1870; Boston, March 10th, 1870; and New York, Saturday March 12, 1870"; and *PLMG,* 17 March 1870.

25. Benedict, *Red Metal,* 111–12; QMC records, *Annual Report for 1874,* 7–8; and Gates, *Michigan Copper,* 48–49.

26. Gates, *Michigan Copper,* 47, 49, 203. The premier mining publication of the time, the *EMJ* has numerous references to the operations of "the Combination," in controlling prices; see *EMJ* 20 (20 November 1875): 509; 20 (27 November 1875): 533; and 21 (1 January 1876): 10. Orris C. Herfindahl and Benedict suggest that C & H directed a pool that functioned without interruption from at least 1876 through 1886; Herfindahl, *Copper Costs and Prices, 1870–1957* (Baltimore: Johns Hopkins University Press, 1959), 247–48, and Benedict, *Red Metal,* 112–13.

27. Gates, *Michigan Copper,* 47–48, 197–98, 203, 230.

28. QMC records, "'Pool' Copper Agreement, Dec. 20, 1882"; ibid., "Copy of 'Pool' Agreement, March 29/83"; Benedict, *Red Metal,* 112–13; and Gates, *Michigan Copper,* 50–51.

29. Gates, *Michigan Copper,* 50–52, 198, 204; and QMC records, Minutes of Directors' Meetings, 19 November 1884.

30. Gates, *Michigan Copper,* 76–78, 198, 204; and Kenneth Ross Toole, "The Anaconda Copper Mining Company: A Price War and a Copper Corner," *Pacific Northwest Quarterly* 41 (October 1950): 318–21.

31. E. Benjamin Andrews, "The Late Great Copper Syndicate," *Quarterly Journal of Economics* 3 (1889): 508–16; and Toole, "The Anaconda Copper Mining Company," 324–25.

32. *Mineral Industry, 1892,* 129; and M. A. Abrams, "The French Copper Syndicate, 1887–1889," *Journal of Economic and Business History* 4 (1931–32): 422–25.

33. Gates, *Michigan Copper,* 80–81; *Commercial and Financial Chronicle* 48 (23 March 1889): 382–83; *EMJ* 67 (23 March 1889): 274; *Mining and Scientific Press* 58 (23 March 1889): 58; and *Mineral Industry, 1892,* 129.

34. Smith, *American Copper Production,* 7, 73–84. Smith offered two explanations: (1) the mine operators believed that a worldwide copper syndicate would obviate the need for tariffs, and (2) they had entered copper and brass manufacturing, and as a result were more concerned about buying the cheapest copper available in the world market. Neither explanation is documented, and neither is very convincing. The larger mine operators, especially in the West, were among the lowest-cost producers in the world, so tariff protection was largely irrelevant to them.

35. Gates, *Michigan Copper,* 108, 208–9; *Mineral Statistics of the State of Michigan for 1880,* 19, 65–66, 78–79, 121–24, 150; *Mines and Mineral Statistics of the State of Michigan for 1891,* 45; and Lankton and Hyde, *Old Reliable,* 152.

36. This theme is developed in more detail in Charles K. Hyde, "Undercover and Underground: Labor Spies and Mine Management in the Early Twentieth Century," *Business History Review* 60 (Spring 1986): 1–27.

37. *PLMG,* 15, 29 July 1865; 19 April 1866; and 14 June 1866. For the managers' efforts to prevent strikes, see the Record Book of the Houghton County Mine Agents Union, QMC records. The mines involved were the Calumet, Hecla, Quincy, Pewabic, Franklin, Adams, Grand Portage, South Pewabic, Agawam, and Isle Royale.

38. *PLMG,* 21, 28 April 1870 and 2 February 1871; and Gates, *Michigan Copper,* 58.

39. John Rowe, *The Hard-Rock Men: Cornish Immigrants and the North American Mining Frontier* (New York: Harper & Row, 1974), 167. Documents relating to the strike include a printed circular, "R. G. Wood, Supt., Office of the Calumet and Hecla Mining Company, Calumet, Mich., April 17, 1872, To the Employees of the Calumet and Hecla Mining Company," and telegram, R. J. Wood to Henry P. Baldwin, Governor of Michigan, 8 May 1872, State of Michigan Archives, Records of the Executive Office, Record Group 44, Box 150, File 1.

40. Bartholomew Shea, Sheriff of Houghton County, Michigan to His Excellency, the Honorable Henry P. Baldwin, Governor of Michigan, telegram, 8 May 1872; and Lieutenant General P. H. Sheridan, Headquarters, Military Division of the Missouri, to H. P. Baldwin, telegram, 11 May 1872, State of Michigan Archives, Records of the Executive Office, Record Group 44, box 150, file 1.

41. *PLMG,* 16 May 1872; and State of Michigan Archives, Record Group 44, box 15, file 1.

42. *PLMG,* 23, 30 May 1872; James O'Grady, Circuit Judge, 12th Judicial Circuit to His Excellency Henry P. Baldwin, Governor of the State of Michigan, 25 May 1872; and W. L. Whitmore, New York Iron Mine, Marquette, Michigan, to H. P. Baldwin, 23 May 1872, both in State of Michigan Archives, Record Group 44, box 150, file 1.

43. A. J. Corey to W. R. Todd, 11 June 1873, 20 December 1873, 2, 7 January 1874, QMC records; *PLMG,* 15 January 1874 and 22 January 1874.

44. Alexander Agassiz to James N. Wright, 5 January 1874, quoted in Gates, *Michigan Copper,* 113–14.

45. Gates, *Michigan Copper,* 114.

46. B. R. Mitchell and Phyllis Deane, *Abstract of British Historical Statistics* (Cambridge: Cambridge University Press, 1962), 159; Joanne Fox Przeworski, *The Decline of the Copper Industry in Chile and the Entrance of North American Capital, 1870–1916* (New York: Arno Press, 1980), 17; William W. Culver and Cornel J. Reinhart, "Capitalist Dreams: Chile's Response to Nineteenth-Century World Copper Competition," *Comparative Studies in Society and History* 31 (October 1989): 726; Mulhall, *Dictionary of Statistics,* 156; and *Mineral Industry, 1892,* 118.

47. *Mineral Industry, 1892,* 116–17.

48. Gates, *Michigan Copper,* 197–98.

49. Robert E. Barclay, *Ducktown Back in Raht's Time* (Chapel Hill: University of North Carolina Press, 1946), 87–97, 155; Clarence S. Ross, *Origin of the Copper Deposits of the Ducktown Type in the Southern Appalachian Region,* U.S. Geological Survey Professional Paper 179 (Washington, D.C.: USGPO, 1935), 95; and M.-L. Quinn, "Industry and Environment in the Appalachian Copper Basin, 1890–1930," *Technology and Culture* 34 (July 1993): 580–85, 604–7.

50. Walter Harvey Weed, "Notes on the Copper Mines of Vermont," U.S. Geological Survey Bulletin No. 225 (Washington, D.C.: USGPO, 1904), 190–99; Collamer M. Abbott, *Green Mountain Copper: The Story of Vermont's Red Metal* (Randolph, Vt.: Herald Printery, 1973), 9–27; U.S. Bureau of the Census, *Ninth Census, 1870,* vol. 3: *Statistics of Wealth and Industry,* 767; Raphael Pumpelly, *Report on the Mining Industries of the United States,* U.S. Bureau of the Census, *Tenth Census, 1880,* vol. 15, pt. 2 (Washington, D.C.: USGPO, 1886), 798; and *Mineral Industry, 1892,* 109.

51. Gates, *Michigan Copper,* 197; Raphael Pumpelly, *Report on the Mining Industries of the United States,* U.S. Bureau of the Census, *Tenth Census, 1880,* vol. 15, pt. 2 (Washington, D.C.: USGPO, 1886), 799–800; *Mineral Industry, 1892,* 108–9; and F. E. Richter, "The Copper-Mining Industry in the United States, 1845–1925," *Quarterly Journal of Economics* 41 (1927): 252–55.

52. Lankton and Hyde, *Old Reliable,* 154; Gates, *Michigan Copper,* 208–10; and Copper Country Commercial Club, *Strike Investigation* (Chicago: Donahue & Company, 1913), 8.

53. Hyde, "Undercover and Underground," 1–27, documents the increased use of spies as a way of gathering information.

54. Thomas F. Mason to Samuel B. Harris, 14 April 1890; Harris to Mason, 16 April 1890; Harris to Mason, 21 June 1890; Harris to William Rogers Todd, 23 June 1890; all in QMC records; *EMJ* 49 (Januaary–June 1890): 503, 713, 719; *EMJ* 50 (July–December 1890): 14, 107; and Allen F. Rees to the Honorable Cyrus G. Luce, Governor, telegrams, 20 June and 23 June 1890, State of Michigan Archives, Records of the Executive Office, Record Group 44, box 150, file 6.

55. The *Engineering and Mining Journal* provided the best overall coverage of strikes between 1890 and 1913, in part because the local newspapers often ignored labor unrest. There is evidence of strikes in 1892 (Osceola and Atlantic);

1893 (Calumet and Hecla); 1894 (Tamarack); 1896 (Quincy); 1897 (Atlantic); 1900 (Quincy); 1904 (Atlantic, Baltic, Quincy, and Trimountain); 1905 (Quincy); and 1906 (Quincy and Michigan). The 1892 unrest is documented in Allen F. Rees to J. S. Farrar, 6 June 1892, and J. S. Farrar to Governor Edward Winans, 9 June 1892, State of Michigan Archives, Records of the Executive Office, Record Group 44, box 150, file 7; and in letters from Thomas Nelson, Treasurer, Osceola Mining Company to John Daniell, Superintendent, 6, 7, 9, 14, 16 June 1892, John Daniell Collection, Huntington Library, San Marino, Calif., box 2, folder 3. The strikes of 1894 and 1896 are discussed in Calumet and Hecla Mining Company, *Summary of the Operations of the Calumet and Hecla Mining Company for the Year Ending April 30, 1894,* n.p.; Lankton and Hyde, *Old Reliable,* 84–85; and Thomas F. Mason to Samuel B. Harris, 30 April 1896, QMC records.

56. Charles Lawton to William R. Todd, 23 July 1906; Lawton to Todd, 4, 10, 11 August 1906; and list of demands under the heading "To the Quincy Mining Company" dated Hancock, Michigan, July 27, 1906, all in the QMC records.

57. Arthur E. Poutinen, *Finnish Radicals and Religion in Midwestern Mining Towns, 1865–1914* (New York: Arno Press, 1979), 203–6; Arthur W. Thurner, *Rebels on the Range: The Michigan Copper Miners' Strike of 1913–1914* (Lake Linden, Mich.: John Forster Press, 1984), 29–31; State of Michigan Archives, Records of the Executive Office, Record Group 44, box 12, file 3, and box 176, file 8.

58. Vernon H. Jensen, *Heritage of Conflict: Labor Relations in the Nonferrous Metals Industry up to 1930* (Ithaca, N.Y.: Cornell University Press, 1950), 288; and Arthur W. Thurner, "The Western Federation of Miners in Two Copper Camps: The Impact of the Michigan Copper Miners' Strike on Butte's Local No. 1," *Montana, the Magazine of Western History* 33 (Spring 1983): 30–45. The only full-length scholarly study is Arthur W. Thurner, *Rebels on the Range: The Michigan Copper Miners' Strike of 1913–1914* (Lake Linden, Mich.: John Forster Press, 1984). Two valuable contemporary studies include a wealth of detailed information on the strike: U.S. Department of Labor, Bureau of Labor Statistics, *Michigan Copper District Strike,* Bulletin No. 139 (Washington, D.C.: USGPO, 1914); and Copper Country Commercial Club of Michigan, *Strike Investigation* (Chicago: M. A. Donahue & Co., 1913).

59. Thurner, *Rebels on the Range,* 35.

60. Ibid., 38–42, 46.

61. Ibid., 43; and Gates, *Michigan Copper,* 129–31, 258nn. 79–81.

62. Copper Country Commercial Club, *Strike Investigation,* 30–42, 77–82.

63. Lankton and Hyde, *Old Reliable,* 110; and Gates, *Michigan Copper,* 131.

64. Thurner, *Rebels on the Range,* 39, 46; U.S. Bureau of Labor Statistics, *Michigan Copper District Strike,* Bulletin No. 139 (Washington, D.C.: USGPO, 1914), 28–29; and Jensen, *Heritage of Conflict,* 275.

65. U.S. Department of Labor, Bureau of Labor Statistics, *Michigan Copper District Strike,* 101–3; and William A. Sullivan, "The 1913 Revolt of the Michigan Copper Miners," *Michigan History* 43 (1959): 309–14.

66. Gates, *Michigan Copper,* 132–34; and Lankton and Hyde, *Old Reliable,* 129.

67. Thurner, *Rebels on the Range,* 191–207, 229, 255–56.

68. Lankton and Hyde, *Old Reliable,* 130–31; and *EMJ* 104 (1917): 277, 321.

69. For copper production by firm for the period 1845–1925, see B. S. Butler

and W. S. Burbank, *The Copper Deposits of Michigan*, U.S. Geological Survey Professional Paper 144 (Washington, D.C.: USGPO, 1929), 76–98.

70. Butler and Burbank, *The Copper Deposits of Michigan*, 76–98; and Gates, *Michigan Copper*, 65.

71. Lankton and Hyde, *Old Reliable*, 54, 101; and Gates, *Michigan Copper*, 70–73. The share of Michigan production in 1904 controlled by the four was as follows: Calumet and Hecla (38.6 percent); the Paine-Stanton group (30.0 percent); the Bigelow-Lewisohn properties (18.4 percent); and the Quincy Mining Company (8.8 percent).

72. Gates, *Michigan Copper*, 122–25.

73. Ibid., 89–90; and "Data on Quincy Shafts," QMC records.

74. Lankton and Hyde, *Old Reliable*, 110–12; Butler and Burbank, *Copper Deposits of Michigan*, 79–97.

75. Butler and Burbank, *Copper Deposits of Michigan*, 94–95; and Lankton and Hyde, *Old Reliable*, 63–64, 109.

76. Lankton and Hyde, *Old Reliable*, 65–66.

77. Lankton and Hyde, *Old Reliable*, 61; and Gates, *Michigan Copper*, 67, 126.

78. Gates, *Michigan Copper*, 68–69; Lankton and Hyde, *Old Reliable*, 62, 76; and C. Harry Benedict, *Red Metal: The Calumet and Hecla Story* (Ann Arbor: University of Michigan Press, 1952), 183–93. Benedict, a metallurgist who spent most of his career with Calumet and Hecla and who developed the ammonia leaching process, described these innovations in greater technical detail in *Lake Superior Milling Practice: A Technical History of a Century of Copper Milling* (Houghton: Michigan College of Mining and Engineering, 1955).

79. James Ralph Finlay, *The Cost of Mining: An Exhibit of the Results of Important Mines throughout the World*, 2d ed. (New York: McGraw-Hill, 1910), 129–204. Finlay's estimate of Calumet and Hecla's costs of 9 cents per pound for 1908 (p. 164) are merely a guess, because C & H did not reveal its costs in published reports. Similarly, his figures for Anaconda (p. 170) are flawed because he fails to take into consideration the silver and gold recovered from the copper ores.

80. Lankton and Hyde, *Old Reliable*, 109–10; Benedict, *Red Metal*, 158; and Gates, *Michigan Copper*, 125.

81. Lankton and Hyde, *Old Reliable*, 110; and Gates, *Michigan Copper*, 126.

82. Lankton and Hyde, *Old Reliable*, 130; and Benedict, *Red Metal*, 159.

83. Gates, *Michigan Copper*, 138–42.

84. H.A.C. Jenison, "Costs of American Copper Production, 1909–1920 Inclusive," *EMJ* 113 (1922): 442–45.

Chapter 4. The Richest Hill on Earth

1. Rodman Wilson Paul, *Mining Frontiers of the Far West, 1848–1880* (New York: Holt, Rinehart and Wilson, 1963); Otis E. Young Jr., *Western Mining: An Informal Account of Precious-Metals Prospecting, Placering, Lode Mining, and Milling on the American Frontier from Spanish Times to 1893* (Norman: University of Oklahoma Press, 1970); Ronald C. Brown, *Hard-Rock Miners: The Intermountain West, 1860–1920* (College Station: Texas A & M University Press, 1979); Richard E. Lingenfelter, *The Hardrock Miners: A History of the Mining Labor Movement in the American*

West, 1863–1893 (Berkeley: University of California Press, 1974); Mark Wyman, *Hard Rock Epic: Western Miners and the Industrial Revolution, 1860–1910* (Berkeley: University of California Press, 1979); and U.S. Bureau of the Census, *Historical Statistics of the United States: Colonial Times to 1970*, pt. 1 (Washington, D.C.: USGPO, 1971), 584.

2. Michael P. Malone, *The Battle for Butte: Mining and Politics on the Northern Frontier, 1864–1906* (Seattle: University of Washington Press, 1981), 10, 15–16; Malone and Richard B. Roeder, *Montana: A History of Two Centuries* (Seattle: University of Washington Press, 1976), 50–55; Ralph I. Smith, *History of the Early Reduction Plants of Butte, Montana* (Butte: Montana School of Mines, 1953), 3–4; and Richard H. Peterson, *The Bonanza Kings: The Social Origins and Business Behavior of Western Mining Entrepreneurs, 1870–1900* (Lincoln: University of Nebraska Press, 1971), 25–26, 145–46.

3. John W. Hakola, "Samuel T. Hauser and the Economic Development of Montana: A Case Study in Nineteenth-Century Frontier Capitalism" (Ph.D. diss., Indiana University, 1961), 102–4; Malone, *Battle for Butte*, 11–12, 16–17, 20–21; and E. G. Leipheimer, *The First National Bank of Butte: Seventy-Five Years of Continuous Banking Operation, 1877 to 1952* (St. Paul, Minn.: Brown & Bigelow, 1952), 9–18. The inventory is in the A. M. Holter Papers, Montana Historical Society, box 117, folder 10.

4. Carl B. Glasscock, *The War of the Copper Kings: Builders of Butte and Wolves of Wall Street* (New York: Bobbs-Merrill, 1935), 59–60; Michael P. Malone, "Midas of the West: The Incredible History of William Andrews Clark," *Montana, The Magazine of Western History* 33 (Autumn 1983): 5–6; William R. Mangum, *The Clarks: An American Phenomenon* (New York: Silver Bow Press, 1941), 46–47; James A. MacKnight, *The Mines of Montana: Their History and Development to Date* (Helena, Mont.: C. K. Wells, 1892), 32, 39, 52; Malone, *Battle for Butte*, 13–17, 22; James E. Fell Jr., *Ores to Metals: The Rocky Mountain Smelting Industry* (Lincoln: University of Nebraska Press, 1979), 51, 140–41; and James High, "William Andrews Clark: Westerner," *Arizona and the West* 2 (Autumn 1960): 247–50.

5. Malone, *Battle for Butte*, 16–20; and Isaac F. Marcosson, *Anaconda* (New York: Dodd, Mead & Co., 1957), 27.

6. Marcosson, *Anaconda*, 31–33.

7. Fremont Older and Cora Older, *George Hearst: California Pioneer* (Los Angeles: Westernlore Press, 1966), 83–110; and Marcosson, *Anaconda*, 34–39.

8. Marcosson, *Anaconda*, 27, 46–47; Malone, *Battle for Butte*, 24–29; and Kenneth Ross Toole, "The Anaconda Copper Mining Company: A Price War and a Copper Corner," *Pacific Northwest Quarterly* 41 (October 1950): 315–17.

9. Malone and Roeder, *Montana*, 131–33; Malone, *Battle for Butte*, 30–31; and Titus Ulke, "Characteristic American Metal Mines: The Anaconda Copper Mine and Works," *Engineering Magazine* 13 (July 1897): 521–24. Ulke visited the Anaconda in 1893.

10. Malone, *Battle for Butte*, 15, 22; and Smith, *History of the Early Reduction Plants*, 8.

11. Robert G. Raymer, *A History of Copper Mining in Montana* (Chicago: Lewis Publishing Company, 1930), 11; M. A. Leeson, ed., *History of Montana, 1739–1885* (Chicago: Warner, Beers & Company, 1885), 950–51; Samuel T. Hauser Papers,

Montana Historical Society, box 60, folder 18; and A. M. Holter Papers, Montana Historical Society, box 117, folders 8, 12.

12. Malone, *Battle for Butte*, 22, 48, 196; and Smith, *History of the Early Reduction Plants*, 8.

13. Gates, *Michigan Copper*, 71; Albert S. Bigelow to John Daniell, 18 April, 28 May, 30 June, 8 July 1887, John Daniell Papers, Huntington Library, San Marino, Calif., box 1, folder 4; Raymer, *History of Copper Mining in Montana*, 17–18; MacKnight, *Mines of Montana*, 46–47; and Agreement between William A. Clark and the Boston and Montana Consolidated Copper and Silver Mining Company, 31 December 1887, ACM Co. Papers, box 169, folder 9.

14. Malone, *Battle for Butte*, 49; Great Falls Tribune, *Great Falls, Montana: Historic and Scenic* (Great Falls: Great Falls Tribune, 1899), 2–4, 20–23; Richard B. Roeder, "A Settlement on the Plains: Paris Gibson and the Building of Great Falls," *Montana: The Magazine of Western History* 42 (Autumn 1992): 6–8; and Albro Martin, *James J. Hill and the Opening of the Northwest* (New York: Oxford University Press, 1976), 333–35, 348–49; Fred L. Quivik, "Power for the Copper Industry: Hydroelectric Developments along the Great Falls of the Missouri River, 1890–1957," paper read at the 17th Annual Meeting of the Society for Industrial Archeology, Wheeling, West Virginia, May 1988; "Agreement, Great Falls Water Power and Townsite Company and the Boston and Montana Consolidated Copper and Silver Mining Company, 12 September 1889," ACM Co. Papers, box 169, folder 14; MacKnight, *Mines of Montana*, p.44; and *Report of the Boston and Montana Consolidated Copper and Silver Mining Company*, for 1894–1898, ACM Co. Papers, printed materials.

15. Raymer, *History of Copper Mining in Montana*, 22; James Arthur MacKnight, *The Mines of Montana: Their History and Development To Date* (Helena, Mont.: C. K. Wells, 1892), 40–42; and Albert S. Bigelow to John Daniell, 24 July 1888, John Daniell Papers, Huntington Library, San Marino, Calif., box 1, folder 5.

16. Raymer, *History of Copper Mining in Montana*, 22, 29–30; Albert S. Bigelow to Charles H. Palmer, 3, 14, 17, 18 November 1890 and 10, 21 March 1891, ACM Co. Papers, box 352, folders 1, 18; and Bigelow to Palmer, 21, 28, 29 November 1893, 16 April, 2 May, 31 August, and 5 November 1894, ACM Co. Papers, box 353, folders 3, 4.

17. Malone, *Battle for Butte*, 45–46; and Raymer, *History of Copper Mining in Montana*, 27–28. Lists of Anaconda stockholders dated 22 November 1898 and 20 April 1899 show J. B. Haggin with 629,990 shares and 613,290 shares respectively out of a total of 1,200,000; ACM Co. Papers, box 38, folder 2.

18. M. Donahoe to Marcus Daly, 11 November 1889,; and Charles F. Adams to J. B. Haggin, 19, 30 November 1889; Marcus Daly to J. B. Haggin, 21 July 1890; and Statement of Losses Sustained by the Anaconda Mining Company, etc., 22 August 1890, ACM Co. Papers, box 131, folders 11, 12.

19. The ACM Co. Papers (box 131, folders 11, 12) include extensive correspondence on this issue between January and October 1891 between Daly, Haggin, Donahoe, Keyser, T. F. Oakes (President of the Northern Pacific Railroad), Sidney Dillon (President of the Union Pacific Railway), J. T. Odell (General Manager of the Baltimore and Ohio Railroad), and others; Malone, *Battle for Butte*, 40; Martin, *James J. Hill*, 346–48; Roeder, "A Settlement on the Plains," 8; Marcosson, *Anaconda*, 53; and *Third Annual Report of the Anaconda Copper Mining Company, For the*

Year Ending 30th June 1898, 20–22. The ACM Co. annual reports cited in this chapter are located in the Library of the Montana School of Mines, Butte.

20. ACM Co. Papers, box 37, folder 3, and box 49, folder 9, contracts, with extensions, for 1891–1901. A series of about three dozen letters dealing with this problem, mainly between officials of the Baltimore company and the Anaconda office in Butte, can be found in ACM Co. Papers, box 3, folder 19; box 5, folder 18; and box 6, folder 1.

21. *Third Annual Report of the Anaconda Copper Mining Company for the Year Ending 30th June 1898*, 12–16; and "Statement of Profit and Loss Account for Three Years Ending June 30th, 1898," ACM Co. Papers, box 48, folder 21.

22. Smith, *History of the Early Reduction Plants*, 14–15.

23. MacKnight, *Mines of Montana*, 22; Harry C. Freeman, *A Brief History of Butte, Montana: The World's Greatest Mining Camp* (Chicago: Henry O. Shepard, 1900), 90; and Robert Grant, Assistant General Manager, Parrot Silver and Copper Company, to A. M. Holter, 6 November 1896, in the A. M. Holter Papers, Montana Historical Society, box 117, folder 8.

24. Fred Quivik, "Montana's Minneapolis Bridge Builders," *IA: The Journal of the Society for Industrial Archeology* 10 (1984): 42; Freeman, *A Brief History of Butte, Montana*, 86–96; Work Projects Administration, Writers' Program, *Copper Camp: Stories of the World's Greatest Mining Town, Butte, Montana* (New York: Hastings House, 1943), 198–202; and William B. Daly et al., "Mining Methods in the Butte District," *Transactions, A.I.M.E.* 72 (1925): 262–63.

25. B. H. Dunshee, "Timbering in the Butte Mines," *Transactions, A.I.M.E.* 46 (1913): 137–45.

26. John Gillie, "The Use of Electricity in Mining in the Butte District," *Transactions, A.I.M.E.* 46 (1913): 817–20; R. E. Wade, "The Electrification of the Butte, Anaconda & Pacific Railway," idem, 820–25; William B. Daly et al., "Mining Methods in the Butte District," *Transactions, A.I.M.E.* 72 (1925): 264–74; and William B. Daly, "Evolution of Mining Practice at Butte," *EMJ* 128 (24 August 1929): 280–81.

27. Frederick Laist, "History of Reverberatory Smelting in Montana, 1879 to 1933," *Transactions, A.I.M.E.* 106 (1933): 25–33.

28. Don MacMillan, "The Butte 'Smoke Messiahs' and Their War against Air Pollution," *The Speculator: A Journal of Butte and Southwest Montana History* 1 (Summer 1984): 50.

29. Duane A. Smith, *Mining America: The Industry and the Environment, 1800–1980* (Lawrence: University of Kansas Press, 1987), 45; and MacMillan, "The Butte 'Smoke Messiahs,'" 50–54.

30. A. M. Holter Papers, Montana Historical Society, box 60, folders 12, 17, 20; T. Egleston, "Bessemerizing Copper Mattes," *School of Mines Quarterly* 6 (May 1885): 320–35; H. O. Hoffman, "Notes on the Metallurgy of Copper of Montana," *Transactions, A.I.M.E.* 34 (1904): 261–62; and F. E. Richter, "The Copper Mining Industry of the United States, 1845–1925," *Quarterly Journal of Economics* 41 (1927): 260.

31. Titus Ulke, *Modern Electrolytic Copper Refining* (New York: John Wiley & Sons, 1907), 1–3, 80; and *Mineral Industry, 1892*, 124; 9 (1900): 223–24; 15 (1906): 192–93.

32. Gates, *Michigan Copper*, 89–90; *Mineral Industry, 1894*, 182; Harry Freeman

Campbell, *A Brief History of Butte, Montana: The World's Greatest Mining Camp* (Chicago: H. O. Shepard Company, 1900), 66–83; and Inspector of Mines of the State of Montana, *Annual Report, December 1900* (Helena, Mont.: State Publishing Company, 1901), 19–20.

33. *Mineral Industry, 1897*, 212; and Ulke, *Modern Electrolytic Copper Refining*, 53.

34. Malone, "Midas of the West," 9. The only detailed history of the interrelations of the major "copper kings" is Carl B. Glasscock, *The War of the Copper Kings: Builders of Butte and Wolves of Wall Street* (Indianapolis: Bobbs-Merrill, 1935), a popular, undocumented study.

35. Kenneth Ross Toole, "The Genesis of the Clark-Daly Feud," *Montana Magazine of History* 1 (April 1951): 26–33; Mangum, *The Clarks: An American Phenomenon*, 48–49; and David Emmons, "The Orange and the Green in Montana: A Reconsideration of the Clark-Daly Feud," *Arizona and the West* 28 (Autumn 1986): 225–45.

36. Freeman, *A Brief History of Butte, Montana*, 41, 45; and Malone, "Midas of the West," 10–14.

37. Gates, *Michigan Copper*, 85–86; and Malone, *Battle for Butte*, 133–36.

38. F. Ernest Richter, "The Amalgamated Copper Company: A Closed Chapter in Corporation Finance," *Quarterly Journal of Economics* 30 (1915–1916): 387–88; Malone, *Battle for Butte*, 137–38; and Thomas W. Lawson, *Frenzied Finance: The Crime of the Amalgamated* (New York: Ridgeway-Thayer Company, 1905).

39. Raymer, *History of Copper Mining in Montana*, 27–28; and Marcosson, *Anaconda*, 94–95.

40. Malone, *Battle for Butte*, 137–39, 156–57.

41. Marcus Daly to M. Donahoe, 16, 18, 20, 23, and 30 December 1899; Marcus Daly to Daly, Donahoe & Greenwood, Bankers, 30 December 1899; and Albert C. Burrage to M. Donahoe, 27 January 1900, ACM Co. Papers, box 3, folder 16.

42. Sarah McNelis, *Copper Kings at War: The Biography of F. Augustus Heinze* (Butte: University of Montana Press, 1968), 51–52.

43. Ibid., 37–42; and Malone, *Battle for Butte*, 144, 160.

44. John D. Leshy, *The Mining Law: A Study in Perpetual Motion* (Washington, D.C.: Resources for the Future, 1987), 95, 171, 289–90; and Malone, *Battle for Butte*, 140–41.

45. McNelis, *Copper Kings at War*, 20, 23–27; and P. A. O'Farrell, *Butte: Its Copper Mines and Copper Kings* (New York: J. A. Rogers, 1899), 33–35.

46. McNelis, *Copper King at War*, 31–33, 52–54, and 86–89.

47. Ibid., 54–56.

48. Ibid., 52, 142–85.

49. Quoted in Malone, *Battle for Butte*, 151.

50. Richter, "The Amalgamated Copper Company," 389–91; and Raymer, *History of Copper Mining in Montana*, 68, 70, 77–83, 88.

51. Marcosson, *Anaconda*, 76.

52. Carrie Johnson, "Electrical Power, Copper, and John D. Ryan," *Montana, The Magazine of Western History* 38 (Fall 1988): 26–28.

53. Malone, *Battle for Butte*, 166–67; and Marcosson, *Anaconda*, 136–45, 163, and 223.

54. Richter, "The Amalgamated Copper Company," 405–7; and Gates, *Michigan Copper*, 87–89.

55. O'Farrell, *Butte,* 47–52. ACM Co. Papers show losses of copper of 32.3 percent for the year 1894 (box 54, folder 17) and losses of 27.8 percent and 27.4 percent for the years ending 30 June 1898 and 30 June 1899 respectively (box 48, folder 15).

56. Laist, "History of Reverberatory Smelting in Montana," 25–29, 39–44; William Kelly and Frederick Laist, "Development of Copper Converting at Butte and Anaconda," *Transactions, A.I.M.E.* 106 (1933): 122–31; H. O. Hofman, "Notes on the Metallurgy of Copper in Montana," *Transactions, A.I.M.E.* 34 (1904): 266; and Fredric Quivik, "The Anaconda Company Smelters: Great Falls and Anaconda," *The Speculator: A Journal of Butte and Southwest Montana History* 1 (Summer 1984): 40.

57. Malone, *Battle for Butte,* 42; Quivik, "The Anaconda Company Smelters," 40–43; and *Report of the Anaconda Mining Company for the Year Ending December 31st, 1905,* n.p.

58. *Report of the Anaconda Copper Mining Company for the Year Ending December 31st, 1908,* n.p.; Gordon Morris Bakken, "Was There Arsenic in the Air? Anaconda versus the Farmers of Deer Lodge Valley," *Montana, The Magazine of Western History* 41 (Summer 1991): 30–41; and Quivik, "The Anaconda Company Smelters," 42–43.

59. Raymer, *History of Copper Mining in Montana,* 92.

60. Richter, "The Amalgamated Copper Company," 394–97.

61. Ibid., 400–403; and Malone, *Battle for Butte,* 206.

62. Vernon H. Jensen, *Heritage of Conflict: Labor Relations in the Nonferrous Metals Industry Up to 1930* (Ithaca, N.Y.: Cornell University Press, 1950), 11–14; and Mark Wyman, *Hard Rock Epic: Western Miners and the Industrial Revolution, 1860–1910* (Berkeley: University of California Press, 1979), 156–58.

63. Wyman, *Hard Rock Epic,* 158–59; and Paul Frisch, "Gibraltar of Unionism: The Development of Butte's Labor Movement, 1878–1900," *The Speculator, A Journal of Butte and Southwest Montana History* 2 (Summer 1985): 12–14.

64. Wyman, *Hard Rock Epic,* 159; Lingenfelter, *Hardrock Miners,* 185–88; and Frisch, "Gibraltar of Unionism," 18–20.

65. Wyman, *Hard Rock Epic,* 164–72; and Lingenfelter, *Hardrock Miners,* 218–19 (the quotation in the text is from p. 219).

66. David M. Emmons, *The Butte Irish: Class and Ethnicity in an American Mining Town, 1875–1925* (Urbana: University of Illinois Press, 1989), 23–24; and Gates, *Michigan Copper,* 108.

67. Emmons, *Butte Irish,* 155, 224, 238–39.

68. Calvert, *Gibraltar,* 72; Raymer, *A History of Copper Mining in Montana,* 84; and Agreement, 4 April 1907, between the Anaconda Mill & Smeltermen's Union, No. 117, Western Federation of Miners and the Anaconda Copper Mining Company, ACM Co. Papers, box 59, folder 9, and box 368, folder 16, for the Great Falls works.

69. Jenson, *Heritage of Conflict,* 293–97; and Emmons, *Butte Irish,* 229.

70. Emmons, *Butte Irish,* 184. Jenson, *Heritage of Conflict,* 298–324, gives detailed account of the wrangling between 1902 and 1912.

71. Emmons, *Butte Irish,* 187, 199–200.

72. Ibid., 107–8, 190–91, 225, 230–33, 241–42.

73. Calvert, *Gibraltar,* 72–78.

74. Ibid., 78–79; and ACM Co. Papers, box 224, folder 3.

75. Calvert, *Gibraltar,* 79, 81–89.

76. Arnon Gutfield, "The Speculator Disaster in 1917: Labor Resurgence at Butte, Montana," *Arizona and the West* 11 (Spring 1969): 31–38.

77. Calvert, *Gibraltar,* 115–25; Arnon Gutfield, *Montana's Agony: Years of War and Hysteria, 1917–1921* (Gainesville: University of Florida Press, 1979); *Mineral Industry, 1921,* 151; and Malone and Roeder, *Montana,* 251.

Chapter 5. The Emergence of Arizona

1. Thomas R. Navin, *Copper Mining and Management* (Tucson: University of Arizona Press, 1978), 50–54.

2. Fayette A. Jones, *New Mexico Mines and Minerals, Being an Epitome of the Early Mining History and Resources of New Mexico Mines* (Santa Fe: New Mexico Printing Company, 1904), 35–38; Billy D. Walker, "Copper Genesis: The Early Years of Santa Rita del Cobre," *New Mexico Histoorical Review* 54 (Spring 1979): 7–11, 14–17; Dan Rose, *The Ancient Mines of Ajo* (Tucson: Mission Publishing Company, 1936), 23–25; Robert Glass Cleland, *A History of Phelps Dodge, 1834–1950* (New York: Knopf, 1952), 229–30; Work Projects Administration, Writers' Program, *Arizona: A State Guide* (New York: Hastings House, 1940), 90; and Robert L. Spude, "Mineral Frontier in Transition: Copper Mining in Arizona, 1880–1885" (M.A. thesis, Arizona State University, 1976), 7–8 (hereafter cited as Spude, M.A. thesis).

3. Spude, M.A. thesis, 13–20, 25–29.

4. Rodman W. Paul, *Mining Frontiers of the Far West, 1848–1880* (New York: Holt Rinehardt, & Winston, 1963), 156–59; WPA, *Arizona: A State Guide,* 88; and Frank J. Tuck, *History of Mining in Arizona* (Phoenix: Arizona Department of Mineral Resources, 1961), 40a.

5. Robert L. Spude, "Mineral Frontier in Transition: Copper Mining in Arizona, 1880–1885," *New Mexico Historical Review* 51 (January 1976): 19–20, 26; and *Mineral Industry, 1893,* 239. For a more detailed analysis of the development of the Arizona copper mines, see Spude, M.A. thesis.

6. Norman Carmichael and John Kiddie, "Development of Mine Transportation in Clifton-Morenci District," *Transactions, A.I.M.E.* 70 (1924): 829.

7. Floyd S. Fierman, "Jewish Pioneering in the Southwest: A Record of the Freudenthal-Lesinsky-Solomon Families," *Arizona and the West* 2 (Spring 1960): 58–60; and Arthur L. Walker, "Recollections of Early Day Mining in Arizona," *Arizona Historical Review* 6 (April 1935): 39.

8. "Articles of Incorporation, Longfellow Mining Company," in the Arizona Corporation Commission Records, filmfile 4.4.186, Archives Division, Arizona Department of Library, Archives, & Public Records; David F. Myrick, *Railroads of Arizona,* vol. 3: *Clifton, Morenci and Metcalf Rails and Copper Mines* (Glendale, Calif.: Trans-Anglo Books, 1984), 16, 20, 24; Walker, "Recollections," 39; Fierman, "Jewish Pioneering in the Southwest," 60; and James Colquhoun, *The Early History of the Clifton-Morenci District* (London: William Clowes and Sons, 1935), 22–33. Colquhoun consistently dates early events at the Longfellow two years earlier than all the other sources. The chronology used here is the one accepted by everyone else.

9. Patrick Hamilton, *The Resources of Arizona,* 3d ed. (San Francisco: A. L. Bancroft & Company, 1884), 193–94; Richard J. Hinton, *Hand-Book to Arizona: Its*

Resources, History, Towns, Mines, Ruins, and Scenery, (1878; reprint, Tucson: Arizona Silhouettes, 1954), 110–11; Arthur Wendt, "The Copper-Ores of the Southwest," *Transactions, A.I.M.E.* 15 (May 1886): 31; and *EMJ* 19 (16 January 1875): 38, *EMJ* 29 (14 February 1880): 121, and *EMJ* 34 (2 September 1882): 122.

10. Douglas, "Historical Sketch of Copper Smelting in the United States," 281–83; Walker, "Recollections," 34; Colquhoun, *Early History of the Clifton-Morenci District,* 36–43; Myrick, *Railroads of Arizona,* 3:31; Wendt, "Copper-Ores of the Southwest," 40–45; *EMJ* 19 (16 January 1875): 38; *EMJ* 34 (2 September 1882): 122; and Spude, M.A. thesis, 83.

11. *EMJ* 29 (14 February 1880): 121; "Articles of Association of the Detroit Copper Mining Company of Arizona," July 8, 1872, Arizona Corporation Commission Records, filmfile 4.4.191, Arizona Department of Library, Archives & Public Records; *Articles of Association and By-Laws, of the Detroit Copper Mining Company of Arizona, Organized March 6, 1873* (Detroit: Daily Post Book and Job Printing Establishment, 1873), Burton Historical Collections, Detroit Public Library; and James H. McClintock, *Arizona: Prehistoric, Aboriginal, Pioneer, Modern* (Chicago: S. J. Clarke Publishing Co., 1916), 2:421.

12. Udo Zindel, "Landscape Evolution in the Clifton-Morenci Mining District, Arizona, 1872–1986" (M.A. thesis, Arizona State University, 1987), 33; Myrick, *Railroads of Arizona,* 3:17–18, 24; *EMJ* 34 (2 September 1882): 121–22; Robert Glass Cleland, *A History of Phelps Dodge,* 82; and John D. Leshy, *The Mining Law: A Study in Perpetual Motion* (Washington, D.C.: Resources for the Future, 1987), 108–11. The General Mining Law of 1872 required that minerals had to "discovered" before a claim could be "located," but the claim could not be patented until at least $100 of "assessment work" (i.e., development work) was completed.

13. Colquhoun, *Early History of the Clifton-Morenci District,* 65–67, 72–77.

14. *EMJ* 29 (14 February 1880): 121; Hamilton, *The Resources of Arizona,* 192; Wendt, "The Copper-Ores of the Southwest," 42; Colquhoun, *Early History of the Clifton-Morenci District,* 46–54; Fierman, "Jewish Pioneering in the Southwest," 63; W. Turrentine Jackson, *The Enterprising Scot: Investors in the American West after 1873* (Edinburgh: Edinburgh University Press, 1968), 168; and Spude, M.A. thesis, 190.

15. Wendt, "Copper-Ores of the Southwest," 40.

16. *EMJ* 29 (21 February 1880): 133.

17. "Narrative of Samuel J. Freudenthal," undated typescript at the Arizona Historical Foundation, Tempe, Ariz., 7; and James Colquhoun, *The History of the Clifton-Morenci Mining District* (London: John Murray, 1924), 13–14.

18. *Reports on the Mines of the Arizona Copper Company Limited,* pp. 32–34, Lewis Douglas Papers, Special Collections, University of Arizona Library, box 24-b, microfilm reel 9; and Dr. James Douglas Jr. Papers, Arizona Historical Society, Tucson, box 2, folder 7; *EMJ* 29 (14 February 1880): 121; and *EMJ* 34 (2 September 1882): 122.

19. Myrick, *Railroads of Arizona,* 3:45; and Colquhoun, *Early History of the Clifton-Morenci District,* 78.

20. *The Arizona Copper Company Limited, Prospectus,* 11th August 1882, and *Reports on the Mines of the Arizona Copper Company Limited,* Lewis Douglas Papers, Special Collections, University of Arizona Library, box 24-b, microfilm reel 9.

21. Henry Lesinsky, *Letters Written by Henry Lesinsky to His Son* (New York, 1924), 30–32; and Jackson, *Enterprising Scot*, 163–66.

22. Myrick, *Railroads of Arizona*, 3:74; Jackson, *Enterprising Scot*, 168; *EMJ* 34 (2 September 1882): 122; Wendt, "Copper-Ores of the Southwest," 42; Colquhoun, *Early History of the Clifton-Morenci District*, 71; and H. H. Langton, *James Douglas: A Memoir* (Toronto: University of Toronto Press, 1940), 73–74.

23. Jackson, *Enterprising Scot*, 162–77; *Mineral Industry, 1892*, 236; and Spude, M.A. thesis, 171.

24. James Douglas Jr., "The Cupola Smelting of Copper in Arizona," in Albert Williams Jr., ed., *The Mineral Resources of the United States, 1883 and 1884* (Washington, D.C.: USGPO, 1885), 409; Wendt, "Copper-Ores of the South-west," 49; Jackson, *Enterprising Scot*, 175; Hamilton, *Resources of Arizona*, 196–98; and Walker, "Recollections," 41.

25. Cleland, *History of Phelps Dodge*, 82–84, 97–98, 112–13; Langton, *James Douglas*, 58–60, 73–74; T. A. Rickard, *A History of American Mining* (New York: McGraw-Hill, 1932), 282; and Navin, *Copper Mining and Management*, 231.

26. James Douglas to Phelps, Dodge Company, May 1881 and October 1881; William Church to Phelps, Dodge Company, 13 October 1882; and James Douglas to Phelps, Dodge Company, 1 November 1882, Dr. James Douglas Jr. Papers, Arizona Historical Society, Tucson, box 2, folders 10, 12, 17, 20.

27. Cleland, *History of Phelps Dodge*, 82; Wendt, "Copper-Ores of the South-west," 49–50; and Hamilton, *Resources of Arizona*, 199.

28. Douglas, "Cupola Smelting of Copper in Arizona," 406–7; Carl Henrich, "The Copper Ore Deposits and Copper Production near Clifton, Arizona," *EMJ* 39 (31 January 1885): 69; Wendt, "Copper-Ores of the Southwest," 49–52; Walker, "Recollections," 43; and *Mineral Industry, 1892*, 236. Carl Henrich, who designed the Detroit Copper Company's rectangular furnace, claimed it had a daily capacity of 80 tons. See Henrich, "The Copper Ore-Deposits and the Copper Production Near Clifton, Arizona," *EMJ* 39 (31 January 1885): 69.

29. McClintock, *Arizona*, 2:415.

30. Arthur L. Walker, "Early-Day Copper Mining in the Globe District of Arizona," *EMJ* 125 (14, 28 April 1928): 605–6, 695; Wendt, "Copper-Ores of the Southwest," 60–61; and Spude, "Mineral Frontier in Transition," 25.

31. Walker, "Early-Day Copper Mining in the Globe District," 607, 695; Hamilton, *Resources of Arizona*, 216; Wendt, "Copper-Ores of the Southwest," 65–66; and *Mineral Industry, 1892*, 236.

32. Walker, "Recollections," 23, 26–28; "The Baltimore Copper Works," *EMJ* 32 (6 August 1881): 87–88; and R. Brent Keyser, "Copper," in *Maryland: Its Resources, Industries, and Institutions* (Baltimore: Board of World's Fair Managers, 1893), 116–18.

33. Cleland, *History of Phelps Dodge*, 238; and Langton, *James Douglas*, 54–55.

34. Herbert V. Young, *Ghosts of Cleopatra Hill: Men and Legends of Old Jerome* (Jerome, Ariz.: Jerome Historical Society, 1964), 29–30; and Hamilton, *Resources of Arizona*, 183, 187.

35. Hamilton, *Resources of Arizona*, 183–87; John Carl Brogden, "The History of Jerome, Arizona" (Master's thesis, University of Arizona, 1952), 16–17; Douglas, "Cupola Smelting of Copper in Arizona," 409–10; Wendt, "Copper-Ores of the Southwest," 68–73; and *Mineral Industry, 1892*, 236.

36. Lynn R. Bailey, *Bisbee: Queen of the Copper Camps* (Tucson: Westernlore Press, 1983), 11–17.

37. Ibid., 18–20; Spude, "Mineral Frontier in Transition," 19; and Annie M. Cox, "History of Bisbee, 1877 to 1937" (M.A. thesis, University of Arizona, 1938), 17.

38. Bailey, *Bisbee,* 21–22; and Douglas, "Cupola Smelting of Copper in Arizona," 397. The compilation of smelters built in Arizona between 1880 and 1885 is from Spude, M.A. thesis, 139–40.

39. Bailey, *Bisbee,* 22–26.

40. Ibid., 26–30; Langton, *James Douglas,* 56–57; "Report of the Copper Queen Mines, Bisbee Arizona," prepared by James Douglas for Professor Benjamin Silliman Jr., 8 February 1881; and *Description of Property of the Copper Queen Mining Company, Bisbee, Cachise (sic) County, Arizona,* April 1881, Lewis Douglas papers, Special Collections, University of Arizona Library, box 24-b, microfilm reel 15.

41. James Douglas, "The Copper Queen Mine, Arizona," *Transactions, A.I.M.E.* 29 (1899): 514; Wendt, "Copper-Ores of the Southwest," 58–59; Hamilton, *Resources of Arizona,* 167–68; Bailey, *Bisbee,* 23; and Robert F. Palmquist, "Obliterate the Law of the Apex: The Case of the Arizona Prince Copper Company v. Copper Queen Mining Company and Its Consequences," paper presented at the 1991 Arizona Historical Convention, Globe, Arizona.

42. Bailey, *Bisbee,* 32–33; and Cox, "History of Bisbee," 39.

43. Bailey, *Bisbee,* 33–35; and Hamilton, *Resources of Arizona,* 167.

44. *Copper Queen Mining Company, New York, Report to Stockholders for the Year Ending April 1, 1884* (New York: V. M. Ramer, 1884), 10; Langton, *James Douglas,* 68–71; and Cleland, *History of Phelps Dodge,* 100–102.

45. Spude, M.A. thesis, 134–42.

46. *Mineral Industry, 1892,* 236.

Chapter 6. Arizona and the Rest of the West

1. *Mineral Industry, 1892,* 108–9; *Mineral Industry, 1901,* 176; *Mineral Industry, 1911,* 149; and *Mineral Industry, 1921,* 151.

2. *Mineral Industry, 1903,* 74–76.

3. John D. Leshy, *The Mining Law: A Study in Perpetual Motion* (Washington, D.C.: Resources for the Future, 1987), 95; and James Douglas, "The Copper Queen Mines and Works, Arizona, U.S.A.," *Transactions of the Institution of Mining and Metallurgy* 22 (1913): 546–47.

4. *Mineral Industry, 1910,* 149; and *Mineral Industry, 1920,* 145, 153–55.

5. W. Turrentine Jackson, *The Enterprising Scot: Investors in the American West after 1873* (Edinburgh: Edinburgh University Press, 1968), 176–82.

6. Spude, M.A. thesis, 195–97 (hereafter cited as Spude, M.A. thesis); and James Colquhoun, *The History of the Clifton-Morenci Mining District* (London: John Murray, 1924), 43.

7. James Colquhoun, *The Story of the Birth of the Porphyry Coppers* (London: William Clowes and Sons, 1933), 13–14; James Colquhoun, *History of the Clifton-Morenci Mining District,* 46–53; William C. Conger, "History of the Clifton-Morenci District," in J. Michael Canty and Michael N. Greeley, eds., *History of Mining in Arizona* (Tucson: Mining Club of the Southwest Foundation, 1987), 100–101; and James M. Patton, *History of Clifton* (Mesa, Ariz.: Publication Services,

1977), 18–23. Detailed monthly figures for the Arizona Copper Company's costs of production, summarized every six months, are found in *Cost Statement,* October 1884–March 1892, and *Cost Statement,* April 1892–March 1899, Arizona Copper Company Papers, Special Collections, University of Arizona, vols. 68, 69. The production and price data are from James B. Tenney, "History of Mining in Arizona," p. 438, MS (1929), Special Collections, University of Arizona.

8. Hugh H. Langton, *James Douglas: A Memoir* (Toronto: University of Toronto Press, 1940), 90–91; Walter R. Bimson, *Louis D. Ricketts: Mining Engineer, Geologist, Banker, Industrialist, and Builder of Arizona* (New York: Newcomen Society, 1949), 13; William E. Dodge to James Douglas, 17 December 1887; William Church to William E. Dodge, 7 October 1895; William E. Dodge to William Church, 14 October 1895; William E. Dodge to James Douglas, 17 October 1895; and miscellaneous letters, Walter Douglas to James Douglas, April 1897 through December 1898, all in the Lewis Douglas Papers, Special Collections, University of Arizona, box 5, folder 4, and box 12-a, folders 4, 8, 9, 10.

9. Colquhoun, *The Birth of the Porphyry Coppers,* 16–20; Waldemar Lindgren, *The Copper Deposits of the Clifton-Morenci District, Arizona,* U.S. Geological Survey, Professional Paper No. 43 (Washington, D.C.: USGPO, 1905), 42; Patton, *History of Clifton,* 24–26; and Conger, "History of the Clifton-Morenci District," 101–2.

10. Forrest R. Rickard, "History of Smelting in Arizona," in Canty and Greeley, eds., *History of Mining in Arizona,* 199; and Lindgren, *Copper Deposits of the Clifton-Morenci District,* 44.

11. Clark C. Spence, *British Investments and the American Mining Frontier, 1860–1901* (Ithaca, N.Y.: Cornell University Press, 1958), 244, 253; Conger, "History of the Clifton-Morenci District," 102; Lindgren, *Copper Deposits of the Clifton-Morenci District,* 34, 46–47; William A. Farish, "Report on the Claims of the Clifton Consolidated Copper Mines of Arizona, Clifton, Arizona, April 11, 1901," and *The Clifton Consolidated Copper Mines of Arizona, Limited, Report of First Annual Meeting of Shareholders, held at Winchester House, December 30th, 1902* (London: F. W. Potter, 1903), both in the New England & Clifton Copper Company Papers, Special Collections, University of Arizona, box 1, folder 4; New England & Clifton Copper Company Papers, box 1, folders 2, 7; box 3, folder 1; and box 7, folder 9; Patton, *History of Clifton,* 53; Hermann Hagedorn, *The Magnate: William Boyce Thompson and His Time, 1869–1930* (New York: Reynal & Hitchcock, 1935), 23–49, 64–74; Conger, "History of the Clifton-Morenci District," 102; and Patton, *History of Clifton,* 55–56.

12. Jackson, *Enterprising Scot,* 182–84; and Patton, *History of Clifton,* 57–58. The original firm of Phelps, Dodge & Company, established in 1834, reorganized as the Phelps Dodge Corporation in 1917.

13. Walker, "Early-Day Copper Mining in the Globe District" *EMJ* 125 (14, 28 April 1928): 607, 695–96; and Thomas A. Rickard, *A History of American Mining* (New York: McGraw-Hill, 1932), 292.

14. Robert H. Ramsey, *Men and Mines of Newmont: A Fifty-Year History* (New York: Farrar, Straus and Giraux, 1973), 28–29, 147–49; and *Mineral Industry, 1918,* 156.

15. Langton, *James Douglas,* 90–91; Russell Wahmann, "A Centennial Commemorative: United Verde Copper Company, 1882–1982," *Journal of Arizona His-*

tory 23 (Autumn 1982): 253–54; and Rickard, *History of American Mining,* 290–91. Several hundred letters between Secretary-Treasurer Eugene M. Jerome, Director Frederick Tritle, and Superintendent Frederick E. Murray that show the efforts to sell the property over the period 1886–88 are found in the United Verde Copper Company Papers, Special Collections, University of Arizona Library.

16. Russell Wahmann, *Verde Valley Railroads* (Cottonwood, Ariz.: Starlight Publishing, 1983), 7–10, 19–23; and Rickard, *History of American Mining,* 291.

17. Wahmann, "United Verde Copper Company,," 254–59; and Horace J. Stevens, *The Copper Handbook* (Houghton, Mich.: The Author, 1902), 3:530–32.

18. Malone, *Battle for Butte,* 82; *Mineral Industry, 1893,* 236: *Mineral Industry, 1899,* 159; *Mineral Industry, 1903,* 76; and *Mineral Industry, 1914,* 159.

19. John Jewett, "The Verde Mining District, Arizona," *EMJ* 72 (10 August 1901): 169–71; and Rickard, *History of American Mining,* 370–77.

20. "Early Furnace Equipment at Bisbee, as shown by Ben Williams' Annual Inventory Statements," box 3, folder 28; and undated agreement between the Copper Queen Consolidated Mining Company and the Société Industrielle at Commercialle Des Métaux De Pris of France, box 6, folder 71; both in the Dr. James Douglas Jr. Papers, Arizona Historical Society, Tucson; and Lynn Robinson Bailey, *Bisbee: Queen of the Copper Camps* (Tucson, Ariz.: Westernlore Press, 1983), 35–36. The quotation is from Langton, *James Douglas,* 87.

21. Bailey, *Bisbee,* 38–40; Langton, *James Douglas,* 96–98; and James Douglas, "The Copper Queen Mines and Works, Arizona, U.S.A.," *Transactions of the Institution of Mining and Metallurgy* 22 (1913): 539–42.

22. Langton, *James Douglas,* 87–88; and Bailey, *Bisbee,* 40–42.

23. Bailey, *Bisbee,* 40–47.

24. Ibid., 43–45; Morris J. Elsing, "The Bisbee Mining District: Past, Present, and Future," *EMJ* 115 (27 January 1923): 181–82; and Isabel Shattuck Fathauser, *Lemuel C. Shattuck: "A Little Mining, a Little Banking, and a Little Beer"* (Tucson: Westernlore Press, 1991), 101–14.

25. Lin B. Feil, "Helvetia: Boom Town of the Santa Ritas," *Journal of Arizona History* 9 (Summer 1968): 77–95; Robert L. Spude, "Swansea, Arizona: The Fortunes and Misfortunes of a Copper Camp," *Journal of Arizona History* 17 (Winter 1976): 375–96; James E. Sherman and Barbara H. Sherman, *Ghost Towns of Arizona* (Norman, Okla.: University of Oklahoma Press, 1969), 44–47, 62–63; and David F. Myrick, *Railroads of Arizona,* vol. 1: *The Southern Roads* (Berkeley, Calif.: Howell-North Books, 1975), 349–60.

26. *Mineral Industry, 1921,* 151.

27. Melody Webb Grauman, "Kennecott: Alaskan Origins of a Copper Empire, 1900–1938," *Western Historical Quarterly* 9 (April 1978): 197–98.

28. Ibid., 198; and Robert A. Stearns, "Alaska's Kennecott Copper & the Kennecott Copper Corporation," *Alaska Journal* 5 (Summer 1975): 131–34. For the life and career of Stephen Birch, see Elizabeth Towe, *Ghosts of Kennecott: The Story of Stephen Birch* (Anchorage, Alaska: The Author, 1990).

29. Isaac F. Marcosson, *Metal Magic: The Story of the American Smelting & Refining Company* (New York: Farrar, Straus and Co., 1949), 91–92; and Stearns, "Alaska's Kennecott Copper," 135–36.

30. Grauman, "Kennecott," 210.

31. Ibid., 203; *Mineral Industry, 1912,* 175; *Mineral Industry, 1915,* 137; and Stearns, "Alaska's Kennecott Copper," 136.

32. Grauman, "Kennecott," 205.

33. *Mineral Industry, 1910,* 149; *Mineral Industry, 1915,* 129, 136; and *Mineral Industry, 1920,* 145.

34. A. B. Parsons, *The Porphyry Coppers* (New York: American Institute of Mining and Metallurgical Engineers, 1933), 3–6, 97–98.

35. Colquhoun, *Birth of the Porphyry Coppers,* 23–29.

36. Parsons, *Porphyry Coppers,* 48–54.

37. Parsons, *Porphyry Coppers,* 64–69; Thomas A. Rickard, *The Utah Copper Enterprise* (San Francisco: Mining and Scientific Press, 1919), 28–29; and *The National Cyclopaedia of American Biography,* current vol. D (New York: James T. White & Company, 1934), 245.

38. Leonard J. Arrington and Gary B. Hansen, *"The Richest Hole on Earth": A History of the Bingham Copper Mine* (Logan: Utah State University Press, 1963), 38–40.

39. Ibid., 41–46.

40. Ibid., 47–56; Parsons, *Porphyry Coppers,* 76; and Bailey, *Old Reliable,* 50–53.

41. Parsons, *Porphyry Coppers,* 77–83.

42. Arrington and Hansen, *"The Richest Hole on Earth,"* 90; and Rickard, *Utah Copper Enterprise,* 9.

43. Russell R. Elliott, *Nevada's Twentieth Century Mining Boom: Tonapah, Goldfield, Ely* (Reno: University of Nevada Press, 1966), 173–80; and Elliott, *Growing Up in a Company Town: A Family in the Copper Camp of McGill, Nevada* (Reno: Nevada Historical Society, 1990), 3.

44. Parsons, *Porphyry Coppers,* 120.

45. Ibid., 121–29; and Elliott, *Nevada's Twentieth Century Mining Boom,* 190.

46. 129–31.

47. Elliott, *Nevada's Twentieth Century Mining Boom,* 50; *Mineral Industry, 1911,* 162; *Mineral Industry, 1917,* 135; and *Mineral Industry, 1920,* 145.

48. Rickard, *History of American Mining,* 293–96; *Mineral Industry, 1911,* 176; and *Mineral Industry, 1920,* 153.

49. *The National Cyclopaedia of American Biography* (New York: James T. White & Company, 1932), 22:123.

50. Rickard, *History of American Mining,* 297, 407–8; William C. Epler and Gary Dillard, *Phelps Dodge: A Copper Centennial, 1881–1981* (Bisbee, Ariz.: Copper Queen Publishing Co., 1981), 164; *Mineral Industry, 1916,* 165; and *Mineral Industry, 1920,* 154.

51. Rickard, *History of American Mining,* 298–99; and William Y. Westervelt to John Annan, London, 5 March 1901; Reports by Alexander Hill, 14 March 1901; and Report of James D. Hague to John B. Ball and Edward Dexter, Receivers of the Ray Copper Mines, Limited, 14 May 1901, James D. Hague Papers, Huntington Library, San Marino, Calif., box 19, folder 108.

52. Rickard, *History of American Mining,* 299; and *Mineral Industry, 1911–1920,* passim.

53. Ira B. Joralemon, "The Ajo Copper-Mining District," *Transactions, A.I.M.E.* 49 (1915): 594–95; A. W. Allen, "Ajo Enterprise of the New Cornelia Copper Company," *EMJ* 133 (2 June 1922): 952–53; and Epler and Dillard, *Phelps Dodge,* 65–66.

54. David A. Walker, *Iron Frontier: The Discovery and Early Development of Minnesota's Three Ranges* (St. Paul: Minnesota Historical Society Press, 1979), 131–38; Epler and Dillard, *Phelps Dodge,* 66–67; Rickard, *History of American Mining,* 278–80; and Allen, "Ajo Enterprise," 953–54, 1003–4.

55. Allen, "Ajo Enterprise," 1001–8; and Epler and Dillard, *Phelps Dodge,* 67.

56. Epler and Dillard, *Phelps Dodge,* 68; *Mineral Industry, 1917,* 148; *Mineral Industry, 1918,* 157; and *Mineral Industry, 1920,* 155.

57. Parsons, *Porphyry Coppers,* 208–12.

58. Ibid., 212–13; *Mineral Industry, 1915,* 129; and *Mineral Industry, 1920,* 145.

59. Parsons, *Porphyry Coppers,* 355–60.

60. Ibid., 369–72, 378–82.

61. Carl A. Allen, "Methods and Economies in Mining," *Transactions, A.I.M.E.* 49 (1915): 392–95; Lucien Eaton, "Seventy-Five Years of Progress in Metal Mining," in A. B. Parsons, ed., *Seventy-Five Years of Progress in the Mineral Industry, 1871–1946* (New York: American Institute of Mining and Metallurgical Engineers, 1947), 62–63; and Parsons, *Porphyry Coppers,* 393–97.

62. Parsons, *Porphyry Coppers,* 6; Navin, *Copper Mining and Management,* 122; *Mineral Industry, 1913,* 127; and *Mineral Industry, 1920,* 145.

63. Vernon H. Jensen, *Heritage of Conflict: Labor Relations in the Nonferrous Metals Industry to 1930* (Ithaca, N.Y.: Cornell University Press, 1950), 355–68; and James W. Byrkit, *Forging the Copper Collar: Arizona's Labor-Management War of 1901–1921* (Tucson: University of Arizona Press, 1982), 27–33.

64. Joseph F. Park, "The 1903 'Mexican Affair' at Clifton," *Journal of Arizona History* 18 (Summer 1977): 121–24, 128–34; U.S. Bureau of the Census, *Twelfth Census, 1900,* vol. 1, pt. 1 (Washington, D.C.: USGPO, 1901), 737; F. Remington Barr, "Integrated Results of Sixty Years' Operation of Phelps Dodge Corporation —Morenci Branch," p. 67, in the Alfred T. Barr Papers, Arizona Historical Society, Tucson; James D. McBride, "Henry S. McCluskey: Workingman's Advocate" (Ph.D. diss., Arizona State University, 1982), 58; and Samuel Clifford Dickenson, "A Sociological Study of the Bisbee-Warren District, Prepared for the Arizona State Bureau of Mines, Tucson, December 31, 1917," p. 54, MS, Special Collections, University of Arizona Library. The size of the census "undercount" of Mexican workers in 1900 and later years is simply unknown.

65. Richard E. Lingenfelter, *The Hardrock Miners: A History of the Mining Labor Movement in the American West, 1863–1893* (Berkeley: University of California Press, 1974), 133, 164–69; *Arizona Silver Belt* 17, no. 17 (21 July 1894): 2; and Park, "The 1903 'Mexican Affair' at Clifton," 139.

66. Park, "The 1903 'Mexican Affair' at Clifton," 142–45; Philip J. Mellinger, *Race and Labor in Western Copper: The Fight for Equality, 1896–1918* (Tucson: University of Arizona Press, 1995), 47–51; Roberta Watt, "History of Morenci, Arizona" (M.A. thesis, University of Arizona, 1956), 57–61; and *Writ of Injunction, The Detroit Copper Mining Company of Arizona v. Maximo Avila et al.,* 13 June 1903, Selim M. Franklin Papers, Special Collections, University of Arizona Library, box 80, folder 4.

67. Jensen, *Heritage of Conflict,* 357–60; Mellinger, *Race and Labor in Western Copper,* 79, 87; James D. McBride, "Gaining a Foothold in the Paradise of Capitalism: The Western Federation of Miners and the Unionization of Bisbee," *Journal of*

Arizona History 23 (Autumn 1982): 301–14; and "The Defeat of Unionism at Bisbee," *EMJ* 81 (24 March 1906): 570–71.

68. Jensen, *Heritage of Conflict*, 363–65; and Mellinger, *Race and Labor in Western Copper*, 146–53.

69. James R. Kluger, *The Clifton-Morenci Strike* (Tucson: University of Arizona Press, 1970), 29–33, 38–40, 45–52.

70. Ibid., 67–71; and Hywell Davies and Joseph S. Myers to George W. P. Hunt, Governor of Arizona, 10 February 1916, Archives Division, Arizona Department of Library, Archives & Public Records, Governor's Office, Governor Hunt, Subject Files: Labor, Clifton, 1915.

71. Byrkit, *Forging the Copper Collar*, 103–10.

72. Ibid., 89–93, 308.

73. Ibid., 118–22, 139–43.

74. Ibid., 146; and Spude, "Swansea, Arizona," 393.

75. John H. Lindquist, "The Jerome Deportation of 1917," *Arizona and the West* 11 (Autumn 1969): 238–40.

76. Andrea Yvette Huginnie, "'Strikitos': Race, Class, and Work in the Arizona Copper Industry, 1870–1920" (Ph. D. diss., Yale University, 1991), 303–4.

77. Ibid., 241–44; and Jensen, *Heritage of Conflict*, 398, 422.

78. Jensen, *Heritage of Conflict*, 17; and Byrkit, *Forging the Copper Collar*, 141, 147, 177.

79. Daphne Overstreet, "On Strike! The 1917 Walkout at Globe, Arizona," *Journal of Arizona History* 18 (Summer 1977): 203–15; and Jensen, *Heritage of Conflict*, 397–98.

80. Huginnie, "'Strikitos,'" 305–10, 320.

81. Byrkit, *Forging the Copper Collar*, 157–59; and Jensen, *Heritage of Conflict*, 400–401.

82. Byrkit, *Forging the Copper Collar*, 162–68.

83. Ibid., 186–93, 204; and Jensen, *Heritage of Conflict*, 406–7.

84. Michael E. Parrish's book *Mexican Workers, Progressives, and Copper: The Failure of Industrial Democracy in Arizona during the Wilson Years* (LaJolla, Calif.: Chicano Research Publications, 1979) presents the most convincing analysis of fate of organized labor in Arizona during World War I. The organization of a typical company union of the 1920s can be seen in Phelps Dodge Corporation, Copper Queen Branch, *Employees Representation Plan* (1921), Special Collections, University of Arizona Library.

85. Elliott, *Nevada's Twentieth-Century Mining Boom*, 251–72.

86. Jensen, *Heritage of Conflict*, 262–63.

87. Helen Zeese Papanikolas, "Life and Labor among the Immigrants of Bingham Canyon," *Utah Historical Quarterly* 33 (Fall 1965): 294–96.

88. Ibid., 296–301.

89. Ibid., 303–6, 312–13.

Chapter 7. From the 1920s to the Vietnam War

1. *Mineral Industry, 1927*, 125; Gates, *Michigan Copper*, 198–99; Raymond F. Mikesell, *The Global Copper Industry: Problems and Prospects* (London: Croom

Helm, 1988), 18–20; Mikesell, *The World Copper Industry: Structure and Economic Analysis* (Baltimore: Johns Hopkins University Press, 1979), 15; and George H. Hildebrand and Garth L. Mangum, *Capital and Labor in American Copper, 1845–1990: Linkages between Product and Labor Markets* (Cambridge, Mass.: Harvard University Press, 1992), 188–89.

2. Navin, *Copper Mining and Management* (Tucson: University of Arizona Press, 1978), 158–83.

3. Mikesell, *World Copper Industry,* 32.

4. "Anaconda Copper," *Fortune* 14 (December 1936): 93–94; and 15 (January 1937): 136–37; Navin, *Copper Mining and Management,* 209–13; and Marcosson, *Anaconda,* 143, 185–86.

5. Navin, *Copper Mining and Management,* 215–17; and Marcosson, *Anaconda,* 287, 297–303, 325–39.

6. Marcosson, *Anaconda,* 267–81; and Navin, *Copper Mining and Management,* 215.

7. *The Anaconda Company Annual Report for the Year Ended December 31, 1956,* 8, 16–19.

8. Cleland, *History of Phelps Dodge,* 220–24; and Navin, *Copper Mining and Management,* 233–34.

9. Cleland, *History of Phelps Dodge,* 210–12, 235–37, 240–41; and Navin, *Copper Mining and Management,* 232–34.

10. W. B. Parsons, *The Porphyry Coppers in 1956* (New York: American Institute of Mining, Metallurgical, and Petroleum Engineers, 1957), 190–98.

11. Cleland, *History of Phelps Dodge,* 244–53, 302–3; and Parsons, *Porphyry Copper in 1956,* 51, 59–66.

12. Hildebrand and Mangum, *Capital and Labor in American Copper,* 81–85.

13. Navin, *Copper Mining and Management,* 263–67.

14. Hildebrand and Mangum, *Capital and Labor in American Copper,* 71–72, 81. The most useful histories of the Guggenheim family are Harvey O'Connor, *The Guggenheims: The Making of an American Dynasty* (New York: Covici, Friede, 1937), and Edwin P. Hoyt Jr., *The Guggenheims and the American Dream* (New York: Funk & Wagnalls, 1967). The only serious study of ASARCO is Isaac Marcosson's company-sponsored history, *Metal Magic: The Story of the American Smelting & Refining Company* (New York: Farrar, Straus, 1949).

15. Navin, *Copper Mining and Management,* 242–45; Hildebrand and Mangum, *Capital and Labor in American Copper,* 71–74; and Marcosson, *Metal Magic,* 64–65.

16. Navin, *Copper Mining and Management,* 161, 248–53; Hildebrand and Mangum, *Capital and Labor in American Copper,* 80–81; Parsons, *Porphyry Coppers of 1956,* 234–43; and Marcosson, *Metal Magic,* 129–30. Marcosson also wrote *Copper Heritage: The Story of Revere Copper and Brass Incorporated* (New York: Dodd, Mead & Company, 1955).

17. Robert H. Ramsey, *Men and Mines of Newmont: A Fifty-Year History* (New York: Farrar, Straus and Giroux, 1973), 28–29, 52, 96–97, 126–27; Navin, *Copper Mining and Management,* 148, 285–95, 297–300, 396; and Hildebrand and Mangum, *Capital and Labor in American Copper,* 165–66.

18. Navin, *Copper Mining and Management,* 273–79; and Francis L. Coleman, *The Northern Rhodesia Copperbelt, 1899–1962: Technological Development up to the*

End of the Central African Federation (Manchester: Manchester University Press, 1971), 48–50.

19. Benedict, *Red Metal,* 130–43, 206; and Navin, *Copper Mining and Management,* 201, 308–9.

20. Marcosson, *Metal Magic,* 50–53; O'Connor, *Guggenheims,* 89–95; and Marvin D. Bernstein, *The Mexican Mining Industry, 1890–1950: A Study of the Interaction of Politics, Economics, and Technology* (Albany: State University of New York Press, 1964), 37–38, 50–51.

21. William Epler and Gary Dillard, *Phelps Dodge: A Copper Centennial, 1881–1981* (Bisbee, Ariz.: Copper Queen Publishing Co., 1981), 111–14; Bernstein, *Mexican Mining Industry,* 59–60, 128; and Cleland, *A History of Phelps Dodge,* 131–35.

22. Bernstein, *Mexican Mining Industry,* 57–59; and Marcosson, *Anaconda,* 251–58. For details of Greene's colorful life and the history of the Cananea operations, see C. L. Sonnichsen, *Colonel Greene and the Copper Skyrocket: The Spectacular Rise and Fall of William Cornell Greene; Copper King, Cattle Baron, and Promoter Extraordinary in Mexico, the American Southwest, and the New York Financial District* (Tucson: University of Arizona Press, 1974).

23. Marcosson, *Anaconda,* 255–56, 264–65; Navin, *Copper Mining and Management,* 214; and Bernstein, *Mexican Mining Industry, 1890–1950,* 129.

24. Charles H. Herner, "Gringo Miners along the Rio Moctezuma," *Journal of Arizona History* 29 (Spring 1988): 55–56, 59, 62; C. L. Sonnichsen, "Pancho Villa and the Cananea Copper Company," *Journal of Arizona History* 20 (Spring 1979): 87–99; and George E. Paulsen, "Reaping the Whirlwind in Chihuahua: The Destruction of the Minas de Corralitos, 1911–1917," *New Mexico Historical Review* 58 (July 1983): 255–61.

25. Bernstein, *Mexican Mining Industry, 1890–1950,* 95–126, 128; and Gates, *Michigan Copper,* 198–201.

26. Parsons, *Porphyry Coppers,* 134–45, 159; Przeworski, *Decline of the Copper Industry in Chile,* 276–79.

27. Parsons, *Porphyry Coppers,* 6, 144–45, 151–56; H. Forster Bain and Thomas T. Read, *Ores and Industry in South America* (New York: Harper, 1934), 220–21; and Clark W. Winton, "The Development Problems of an Export Economy: The Case of Chile and Copper," in Markos Mamalakis and Clark W. Winton, eds., *Essays on the Chilean Economy* (Homewood, Ill.: R. D. Irwin, 1965), 216–17.

28. O'Connor, *Guggenheims,* 346; and Parsons, *Porphyry Coppers,* 259–65, 280.

29. Parsons, *Porphyry Coppers,* 6, 273–79; O'Connor, *Guggenheims,* 347–49, 413–14; Marcosson, *Anaconda,* 200–201; and Reynolds, "The Development Problems of an Export Economy," 218.

30. Parsons, *Porphyry Coppers,* 6, 320–34; and Marcosson, *Anaconda,* 211–14.

31. Reynolds, "The Development Problems of an Export Economy," 228; Leland R. Pederson, *The Mining Industry of the Norte Chico, Chile* (Evanston, Ill.: Northwestern University, 1966), 235–46; Marcosson, *Anaconda,* 214–17; and Navin, *Copper Mining and Management,* 218–19.

32. Navin, *Copper Mining and Management,* 124, 310–12, and Elizabeth Dore, *The Peruvian Mining Industry: Growth, Stagnation, and Crisis* (Boulder, Colo.: Westview Press, 1988), 88–92, 102.

33. Marcosson, *Metal Magic,* 170–71; Charles T. Goodsell, *American Corpora-*

tions and Peruvian Politics (Cambridge, Mass.: Harvard University Press, 1974), 49–50, 154–55; Raymond F. Mikesell, *Foreign Investment in Copper Mining: Case Studies of Mines in Peru and Papua New Guinea* (Baltimore: Johns Hopkins Press, 1975), 39–50; and Navin, *Copper Mining and Management*, 236, 249.

34. Simon Cunningham, *The Copper Industry in Zambia: Foreign Mining Companies in a Developing Country* (New York: Praeger, 1981), 64–74; and Navin, *Copper Mining and Management*, 348–55. For the early development of the Northern Rhodesia Copper Belt, see Lucy Pope Cullen, *Beyond the Smoke That Thunders* (New York: Oxford University Press, 1940); Reginald J. B. Moore, *These African Copper Miners: A Study of the Industrial Revolution in Northern Rhodesia, With Principal Reference to the Copper Mining Industry* (London: Livingston Press, 1948); and Kenneth Bradley, *Copper Venture: The Discovery and Development of Roan Antelope and Mufulira* (London: Mufulira Copper Mines, 1952).

35. Orris C. Herfindahl, *Copper Costs and Prices, 1870–1957* (Baltimore: Published for Resources for the Future by Johns Hopkins Press, 1959), 203–4; Ferdinand E. Banks, *The World Copper Market: An Economic Analysis* (Cambridge, Mass.: Ballinger Publishing Co., 1974), 9; and Navin, *Copper Mining and Management*, 84.

36. Hildebrand and Mangum, *Capital and Labor in American Copper*, 102; Navin, *Copper Mining and Management*, 400–401; and Gates, *Michigan Copper*, 198–99.

37. *Mineral Industry, 1895*, 261; *Mineral Industry, 1921*, 148; *Mineral Industry, 1931*, 148–49; and Hildebrand and Mangum, *Capital and Labor in American Copper*, 94–97.

38. Alan R. Raucher, *Public Relations and Business, 1900–1929* (Baltimore: Johns Hopkins Press, 1968), 98–99; and Ray Eldon Hiebert, *Courtier to the Crowd: The Story of Ivy Lee and the Development of Public Relations* (Ames: Iowa State University Press, 1966), 169–71.

39. Navin, *Copper Mining and Management*, 151–53.

40. Ibid., 4; and U.S. Bureau of the Census, *Historical Statistics of the United States: Colonial Times to 1970*, pt. 1 (Washington, D.C.: USGPO, 1975), 602. The copper industry also recycled an even larger quantity of "new scrap," such as shavings and rejected products. New scrap is created in the copper and brass fabricating industries.

41. Mikesell, *World Copper Industry*, 7–10.

42. Hildebrand and Mangum, *Capital and Labor in American Copper*, 105; and Navin, *Copper Mining and Management*, 400–402.

43. Hildebrand and Mangum, *Capital and Labor in American Copper*, 103–5, 108–10.

44. Hildebrand and Mangum, *Capital and Labor in American Copper*, 44–49; and Herfindahl, *Copper Costs and Prices*, 162.

45. Herfindahl, *Copper Costs and Prices*, 92–106.

46. Navin, *Copper Mining and Management*, 125–31; and Gates, *Michigan Copper*, 199, 205.

47. Navin, *Copper Mining and Management*, 133–40; and Gates, *Michigan Copper*, 174, 198–200, 205. Between 1941 and 1945, roughly three-quarters of new copper was sold without subsidy, while most of the rest was sold at prices between 11 and 17 cents a pound. A handful of copper producers received up to 27 cents a pound for their metal (Navin, *Copper Mining and Management*, 140).

48. Herfindahl, *Copper Costs and Prices,* 110–25.

49. Navin, *Copper Mining and Management,* 140–45, 401; and Herfindahl, *Copper Costs and Prices,* 128–29.

50. Hildebrand and Mangum, *Capital and Labor in American Copper,* 172; Navin, *Copper Mining and Management,* 401; and Herfindahl, *Copper Costs and Prices,* 130.

51. President's Materials Policy Commission, *Resources for Freedom,* vol. 2: *The Outlook for Key Commodities* (Washington, D.C.: USGPO, 1952), 33–38, 143–45; Navin, *Copper Mining and Management,* 407; and Mikesell, *World Copper Industry,* 125, 134.

52. The properties, companies, and projected annual capacities were as follows: San Manuel, Arizona (Magma)—70,000 tons; Bisbee, Arizona (Phelps Dodge)—38,000 tons; White Pine, Michigan (Copper Range)—35,000 tons; Miami, Arizona (Miami)—34,000 tons; Yerington, Nevada (Anaconda)—33,000 tons; and Silver Bell, Arizona (ASARCO)—18,000 tons. See Navin, *Copper Mining and Management,* 147–49, 299–300, 401.

53. Navin, *Copper Mining and Management,* 148, 299–300; Herfindahl, *Copper Costs and Prices,* 130, 136–39, 143–44; and A. B. Parsons, *Porphyry Coppers in 1956,* 11.

54. Navin, *Copper Mining and Management,* 155, 159–63.

55. Ibid., 139–41, 154–56, 158–60.

56. The Solow article is quoted at length in Navin, *Copper Mining and Management,* 217–18. Isaac Marcosson's book *Anaconda* (New York: Dodd, Mead & Company, 1957) was the result. Marcosson's books on Anaconda and other copper companies are reliable compilations of factual details and are useful chronologies. They are used in this book as reference works. Marcosson's evaluations of corporate leaders, labor relations, and the like are thoroughly biased and are not used here.

57. Navin, *Copper Mining and Management,* 22.

58. Jensen, *Heritage of Conflict: Labor Relations in the Nonferrous Metals Industry Up To 1930,* 452–66.

59. Vernon H. Jensen, *Nonferrous Metals Industry Unionism, 1932–1954: A Story of Leadership Controversy* (Ithaca, New York: Cornell University Press, 1954) and *Collective Bargaining in the Nonferrous Metals Industry* (Berkeley: University of California Press, 1955).

60. Jensen, *Collective Bargaining in the Nonferrous Metals Industry,* 16–19.

61. Ibid., 21–22, 31–34; Jensen, *Nonferrous Metals Industry Unionism, 1932–1954,* 38, 43; and Gates, *Michigan Copper,* 171–72.

62. Jensen, *Collective Bargaining in the Nonferrous Metals Industry,* 50–54; Hildebrand and Mangum, *Capital and Labor in American Copper,* 146–47; and Jensen, *Nonferrous Metals Industry Unionism, 1932–1954,* 153–280.

63. Hildebrand and Mangum, *Capital and Labor in American Copper,* 210–14, 221–25.

Chapter 8. A Quarter Century of Adjustment and Decline

1. Hildebrand and Mangum, *Capital and Labor in American Copper,* 1, 8, 188–89, 254; and Mikesell, *Global Copper Industry,* 18–20.

2. Navin, *Copper Mining and Management,* 179–83.

3. Ibid., 22, 173; and Hildebrand and Mangum, *Capital and Labor in American Copper,* 179.

4. Navin, *Copper Mining and Management,* 181, 251, 281; and Hildebrand and Mangum, *Capital and Labor in American Copper,* 177.

5. Mark Bostock and Charles Harvey, eds., *Economic Independence and Zambian Copper: A Case Study of Foreign Investment* (New York: Praeger, 1972), 145–77; Simon Cunningham, *The Copper Industry in Zambia: Foreign Mining Companies in a Developing Country* (New York: Praeger, 1981), 264–75; and Navin, *Copper Mining and Management,* 180–81, 281, 357–61.

6. George M. Ingram, *Expropriation of U.S. Property in South America: Nationalization of Oil and Copper Companies in Peru, Bolivia, and Chile* (New York: Praeger, 1974), 72–74, 79–80. 94–95; and Elizabeth Dore, *The Peruvian Mining Industry: Growth, Stagnation, and Crisis* (Boulder, Colo.: Westview Press, 1988), 143–45, 164–71.

7. Ingram, *Expropriation of U.S. Property in South America,* 73, 79–80; Dore, *Peruvian Mining Industry,* 144, 173; Raymond F. Mikesell, *Foreign Investment in Copper Mining: Case Studies of Mines in Peru and Papua New Guinea* (Baltimore: Johns Hopkins Press, 1975), 51–54; and Navin, *Copper Mining and Management,* 181.

8. Ingram, *Expropriation of U.S. Property in South America,* 231–52.

9. Ibid., 253–59.

10. Ibid., 258, 267–68; and Navin, *Copper Mining and Management,* 221–22.

11. Ingram, *Expropriation of U.S. Property in South America,* 270–90, 311–13; Hildebrand, *Capital and Labor in American Copper,* 150–51; and Navin, *Copper Mining and Management,* 180, 221–24.

12. Hildebrand and Mangum, *Capital and Labor in American Copper,* 179.

13. Navin, *Copper Mining and Management,* 210, 309–10.

14. Ibid., 222–25, 281–82; and Hildebrand and Mangum, *Capital and Labor in American Copper,* 151–52.

15. Hildebrand and Mangum, *Capital and Labor in American Copper,* 47, 55–57, 158–61; Navin, *Copper Mining and Management,* 237–39; Epler and Dillard, *Phelps Dodge,* 177–82; and Barbara Kingsolver, *Holding the Line: Women in the Great Arizona Mine Strike of 1983* (Ithaca, N.Y.: ILR Press, 1989), 168–69, 191.

16. Navin, *Copper Mining and Management,* 266.

17. Ibid., 268–72.

18. Hildebrand and Mangum, *Capital and Labor in American Copper,* 46–47, 156–58.

19. Ibid., 44–48, 161–62; and Navin, *Copper Mining and Management,* 251–52.

20. Hildebrand and Mangum, *Capital and Labor in American Copper,* 165–66.

21. Ibid., 168; and Navin, *Copper Mining and Management,* 280–83.

22. Hildebrand and Mangum, *Capital and Labor in American Copper,* 168–69; and Navin, *Copper Mining and Management,* 177n. 4, 321–22.

23. Hildebrand and Mangum, *Capital and Labor in American Copper,* 44–47, 152, 162–64; and Navin, *Copper Mining and Management,* 315–20.

24. Mikesell, *Global Copper Industry,* 18–27.

25. Ibid., 37–42; and Navin, *Copper Mining and Management,* 167.

26. Hildebrand and Mangum, *Capital and Labor in American Copper,* 188–89; and Mikesell, *Global Copper Industry,* 24.

27. Hildebrand and Mangum, *Capital and Labor in American Copper,* 202–3, 252–55.

28. Ibid., 230–43.

29. Ibid., 245–51.

30. Ibid., 203, 252–59.

31. Ibid., 259–63. An excellent detailed account of the strike from the workers' standpoint is Barbara Kingsolver's book *Holding the Line: Women in the Great Arizona Mine Strike of 1983* (Ithaca, N.Y.: ILR Press, 1989).

32. The best study of the copper industry's environmental impact is Duane A. Smith, *Mining America: The Industry and the Environment, 1800–1980* (Lawrence: University Press of Kansas, 1987). The long-term effects of copper smelters are outlined in M.-L. Quinn, "Early Smelter Sites: A Neglected Chapter in the History and Geography of Acid Rain in the United States," *Atmospheric Environment* 23 (1989): 1281–92.

33. Donald MacMillan, "A History of the Struggle to Abate Air Pollution from the Copper Smelters of the Far West, 1885–1933" (Ph.D. diss., University of Montana, 1973), passim; and "The Butte 'Smoke Messiahs' and Their War against Air Pollution," *The Speculator: A Journal of Butte and Southwest Montana History* 1 (Summer 1984): 48–53; Smith, *Mining America,* 74–104; M.-L. Quinn, "Industry and Environment in the Appalachian Copper Basin, 1890–1930," *Technology and Culture* 34 (July 1993): 575–612; and John E. Lamborn and Charles S. Peterson, "The Substance of the Land: Agriculture v. Industry in the Smelter Cases of 1904 and 1906," *Utah Historical Quarterly* 53 (Fall 1985): 308–25.

34. C. E. Mills, "Ground Movement and Subsidence at the United Verde Mine," *Transactions, A.I.M.E.* 109 (1934): 153–72; Richard Rohe, "Man and the Land: Mining's Impact on the Far West," *Arizona and the West* 28 (Winter 1986): 299–338; Richard V. Francaviglia, "Copper Mining and Landscape Evolution: A Century of Change in the Warren Mining District, Arizona," *Journal of Arizona History* 23 (Autumn 1982): 267–98; and Francaviglia, *Hard Places: Reading the Landscape of America's Historic Mining Districts* (Iowa City: University of Iowa Press, 1991).

35. Robert B. Gordon, Tjalling C. Koopmans, William D. Nordhaus, and Brian J. Skinner, *Toward a New Iron Age? Quantitative Modeling of Resource Exhaustion* (Cambridge, Mass.: Harvard University Press, 1987), esp. 123–26, 148–52.

Glossary

This glossary is derived from Horace J. Stevens, ed., *The Copper Handbook* (Houghton, Mich.: Horace J. Stevens, 1904), 4:72–96.

adit. A horizontal mine opening or tunnel used to provide natural drainage and easy removal of rock and ore.
amygdaloid rock. A trap rock containing little pits or amygdules that frequently contain native copper.
apex. That part of an ore vein at or nearest the surface.
barrel work. Copper in small masses detached from its rock matrices and shipped to the smelter in barrels.
blister copper. Copper of 96 to 99 percent purity.
carbonate ore. An ore of any metal with which carbon and oxygen are chemically united.
churn drill. A drill having a churning motion, used for boring test holes.
conglomerate rock. A pudding stone rock formed by the deposit of rock particles on old seabeds that were later covered by other rock strata.
crosscut. An opening similar to a drift except that the opening is set at approximately a right angle to the formation.
diamond drill. A machine for boring holes in rock and which uses black diamonds to form the cutting edges of its bit.
drift. A horizontal opening in a mine, following the direction of the lode or vein.
electrolytic copper. Copper gained from impure metal by electrical decomposition and redisposition in pure form at the opposite pole of the battery.
fathom. Six feet. In stoping, a fathom is a cube six feet on a side.
float copper. Drift copper.
gangue. The particles of foreign rock matter adhering to disseminated ores or native metal. Gangue rock is mechanically, not chemically, united with the ore or metal.
inclined shaft. A shaft sunk at any angle with the horizontal of less than 90 degrees.
ingot copper. A mass of copper cast in a mold.
jig (jigger). A machine for concentrating ore or minerals by means of oscillatory or vibrating motion, aided by water, with separation of the ore from its gangue being effected by the greater specific gravity of the ore.
kibble. A bucket used for hoisting material in a shaft.
man-car. A skip-truck having tiers of seats and used for carrying miners to and from mines operating inclined shafts.
man-engine. A device for raising and lowering miners in deep inclined shafts. It

consists of two long beams working in counterbalance and having platforms at regular intervals.

mass copper. A solid chunk of native metal.

matte copper. Regulus. A product between copper ore and blister copper, varying greatly in the percentage of metal it contains. It is obtained by roasting out sulphur from sulphide copper ores.

mineral. In the Lake Superior district, the native copper, with its adhering gangue of amygdaloid or conglomerate rock, as it comes from the stamp mill and before it goes to the smelter.

native copper. A virgin metal, not an ore.

ore. A chemical union of one or more metallic elements with other elements, usually nonmetallic, of which oxygen, sulphur, and carbon are the most frequent.

pare. A Cornish term for a gang or team of miners.

raise. A shaft or winze that is being opened from below.

refining. The elimination of impurities from crude metals or the separation of metallic alloys obtained in the reduction of ores.

regulus. Matte copper.

reverberatory furnace. A smelting furnace in which the flame from the grate is reflected back on the charge of ore by the roof.

roasting. Driving off sulphur and other volatile elements from ore by heating.

roasting furnace. An oven for the expulsion of sulphur, arsenic, and other volatile elements from ore.

rock house. A building where copper-bearing rock is received and put through crushers before shipment to the stamp mill.

skip. An iron box, open at the top, running on four wheels, and hauled by a cable, used in inclined shafts for hoisting ore and rock.

skip-road or skip-track. A track of T-rails spiked to wooden sleepers, on which the skip runs.

slimes. Exceedingly small particles of rock and mineral held in solution in water, making a slimy mixture.

smelter. Works where ores or crude metals are freed from gangue or chemically united elements by heat.

smelting. The reduction of ores and crude metals in furnaces by heat, with fuel and fluxing material being added to the material being smelted.

square sets. A form of mine timbering with mortise and tenoned sill, top piece, and uprights of equal length, joined at right angles.

stamp mill. A mill for crushing and concentrating minerals.

stamp rock. Rock containing fine copper that can be secured by stamping.

stamps. Machines to crush rock or ore by heavy blows.

steam stamp. A stamp actuated by steam.

stope. The excavation above a drift, or the pay rock remaining unmined above a drift.

stoping. Breaking down the mass of pay rock or ore above a drift.

sulphate ore. An ore of any metal with which sulphur and water are chemically united.

sulphide ore. An ore of any metal with which sulphur is chemically united. Sometimes called a sulphuret.

tailings. Refuse matter from a stamp mill. Also called stamp sands.

tram. To load rock or ore in tram-cars and push it to the shaft.

tram-car. A car running underground on light T-rails, used for carrying rock from the stopes and other workings to the shafts.

trammers. Men who load and tram the broken rock underground.

tribute. The royalty or percentage paid by workmen to owners for the privilege of working a mine.

tutwork. Development work.

vein. A mineral body having defined walls.

vertical shaft. A shaft sunk at an angle of 90 degrees with the horizon, or directly downward toward the center of the earth.

water-jacket furnace. A smelting furnace in which water is circulated between an outer jacket and the furnace proper.

Bibliographic Essay

My research for this study relied heavily on the manuscript collections created by the individuals and firms involved in the copper industry. The most important of these collections are listed below. My research began at archives and libraries on the Atlantic Coast and proceeded westward, much like the copper industry.

The Baker Library at the Harvard University School of Business Administration holds a large collection of mining company annual reports and the R. G. Dun & Company papers, which included credit ratings on copper producers all across the United States.

The Maryland Historical Society archives in Baltimore hold a rich collection of materials on the antebellum eastern copper industry. The most valuable for my research were the Hollingsworth Papers, the Patterson Papers, and the Tyson Record Book.

The Copper Country Historical Collection (CCHC) at Michigan Technological University in Houghton contains a daunting volume of material on the Michigan copper industry. The most notable are large manuscript collections created by the three largest producers: the Calumet and Hecla Mining Company, the Copper Range Consolidated Company, and the Quincy Mining Company. Each of the three large collections contain material relating to scores of other copper mines the three giants absorbed over the years. The CCHC also has the records of more than a score of smaller producers and the largest collection of annual reports generated by Michigan-based copper companies found in any public archive. The Bentley Historical Library at the University of Michigan holds a about a dozen smaller but important collections relating to the early development of the Michigan mines. Finally, the State Archives of Michigan have records relating to mining company incorporation, the Michigan state militia, and the governor's office.

Archival materials on the Montana copper industry are found mainly at the Montana Historical Society archives in Helena. The Anaconda Copper Mining (ACM) Company collection includes more than 500 manuscript boxes, approximately 400 bound volumes, and thousands of photographs, maps, and other materials. The records of the other large Montana producers, including the Butte and Boston Consolidated Mining Company and the Boston and Montana Consolidated Copper and Silver Mining Company, are part of the ACM Company collection. The archives also house the papers of A. M. Holter and Samuel T. Hauser, two early promoters of Montana silver and copper. The library of the Montana School of Mines in Butte has an outstanding collection of annual reports produced by Montana copper companies.

Similarly, a large volume of manuscript material relating the Arizona copper industry has survived. The Special Collections department of the University of

Arizona Library in Tucson houses the richest assortment of primary sources. The most notable are the papers of the Arizona Copper Company, the Detroit Copper Mining Company, the New England & Clifton Copper Company, the Phelps Dodge Corporation, and the United Verde Copper Company. The Lewis Douglas Papers and the Selim M. Franklin Papers are also important sources found there. The Arizona Historical Society archives in Tucson hold the Dr. James Douglas Jr. Papers and the Alfred T. Barr Papers. The papers of Governor George W. P. Hunt are held by the Archives Division, Arizona Department of Library, Archives & Public Affairs in Phoenix.

Finally, the Huntington Library in San Marino, California, has a valuable collection of materials relating to the copper industry, including the John Daniell Papers and the James D. Hague Papers.

These are merely the largest manuscript collections found in the most important repositories. A close reading of the endnotes will reveal scores of other "minor" sources as well.

Printed materials produced by mining engineers in the nineteenth and early twentieth centuries have proven to be invaluable sources. The *Engineering and Mining Journal* (1869–) and the *Transactions* of the American Institute of Mining Engineers (1871–) contain a wealth of technical, geological, and economic details about the copper industry both nationally and internationally.

Richard P. Rothwell, editor of the *Engineering and Mining Journal* (*EMJ*), edited and published an annual statistical supplement to the *EMJ* beginning in 1893 entitled *The Mineral Industry: Its Statistics, Technology and Trade, in the United States and Other Countries from the Earliest Times to the End of 1892* (New York: Scientific Book Publishing Company, 1893). Richard Rothwell edited this valuable compilation covering the years 1892 to 1900, followed by Joseph Struthers (1901–02), D. H. Newland (1903), Edward K. Judd (1904), Walter Renton Ingalls (1905–9), Albert Hill Fay (1910), and Charles Of (1911–12). Finally, G. A. Roush edited *The Mineral Industry* for the years from 1913 to 1941, the final year of publication. Each issue presented a running ten-year summary of output and prices for American copper producers and the world copper industry. Each number also presented a detailed analysis of the major mining districts, including the performance of the individual firms.

The author drew upon a wide variety of other sources as well, including local newspapers, public documents produced by the federal and state governments, and travel narratives. A careful reading of the notes will reveal the rich secondary literature that was vital to the preparation of this work. Countless histories of individual mining districts, communities, and companies were consulted along the way.

Index

About the Author

Charles K. Hyde received his Ph.D. in economic history from the University of Wisconsin–Madison in 1971. His published works include *Technological Change and the British Iron Industry, 1700–1870; Old Reliable: An Illustrated History of the Quincy Mining Company* (with Larry Lankton); *Detroit: An Industrial History Guide; The Northern Lights: Lighthouses of the Upper Great Lakes; Historic Highway Bridges of Michigan;* and numerous articles in scholarly journals. He has taught history at Wayne State University in Detroit since 1974, and he resides in Royal Oak, Michigan.